D1827268

Isoenzyme Analysis

Analytical Sciences Monographs

Isoenzyme Analysis

D. W. Moss

Department of Chemical Pathology
Royal Postgraduate Medical School
(University of London)
Hammersmith Hospital
London W12 0HS

The Chemical Society
Burlington House
London W1V 0BN
1979

British Library Cataloguing in Publication Data

Moss, Donald William
   Isoenzyme analysis. – (Analytical sciences monographs; no. 6   ISSN 0583-8894).
   1. Isoenzymes – Analysis
   I. Title   II. Series
   574.1'925   QP601
   ISBN 0-85186-800-2

Copyright © 1979
The Chemical Society

All Rights Reserved
No part of this book may be reproduced or transmitted
in any form or by any means—graphic, electronic,
including photocopying, recording, taping or information
storage and retrieval systems—without written permission
from The Chemical Society

Printed in Great Britain
by Spottiswoode Ballantyne Ltd. Colchester and London

Contents

Preface

The present recognition that enzymes frequently exist in multiple molecular forms has resulted from the application of a wide variety of experimental techniques. This monograph attempts to draw together the most important of these techniques and to indicate their advantages and limitations in isoenzyme studies. Those which have been developed primarily for, and find their main applications in, other or wider fields of research have not been described in detail; on the other hand, more detailed attention has been given to principles and methods of analysis, such as separation procedures, which have played a particularly important part in isoenzyme investigations. Examples have been drawn most frequently from isoenzyme systems which occur in human tissues, since the clinical and genetic implications of the existence of isoenzymes have provided much of the impetus for the development of appropriate analytical techniques, although the techniques themselves are of general applicability. I hope that this approach will provide those entering the field of isoenzyme research with guidance in their selection of methods, together with access to sources of experimental detail.

The term 'isoenzymes' has been used throughout as a general description of multiple enzyme forms, except when current knowledge compels the exclusion of particular forms from the more restricted definition of this term. Similarly, enzymes are designated by their trivial names, although the identifying numbers allotted by the Enzyme Commission of the International Union of Biochemistry have also been given for enzymes which are referred to more extensively, usually when first mentioned.

I am grateful to many colleagues for helpful discussions. In particular, I thank those authors, editors, and publishers noted in the legends for permission to reproduce illustrations.

D. W. MOSS
London
January 1979

1

Multiple Forms of Enzymes

Definitions and Nomenclature

Enzymes have traditionally been classified by their function; that is, by the nature of the substrate on which they act and the type of reaction which they catalyse. This system of classification establishes various degrees of functional similarity as the basis for the six main groups of enzymes and their constituent sub-classes and sub-sub-classes, to which individual enzymes are assigned according to the scheme proposed originally by the Enzyme Commission of the International Union of Biochemistry. Classification of enzymes by their catalytic specificity was virtually the only feasible system in the early days of the systematic study of enzymes, when information as to the nature of these biocatalysts was almost entirely lacking, and it remains the most generally useful scheme today. However, now that advances in analytical techniques have made it possible to submit individual enzymes to closer and closer scrutiny, it is often found that two or more enzymes may be subsumed under a particular identifying number in the Enzyme Commission's list which functionally are closely similar, but which can be distinguished by other, structurally based criteria. That different molecular forms of an enzyme may occupy the same place in this classification scheme is analogous in some respects to the grouping of isotopes of an element at a particular point in the periodic table, since both classifications are based on similarities of behaviour.

Enzymes are protein molecules of large molecular weight, often with additional, non-protein components, each molecule possessing one or a few small, specialised catalytic sites or active centres at which attachment and transformation of substrate molecules occurs. It may be assumed, therefore, that considerable variation can occur in regions of the enzyme molecule beyond the limits of the active centre, without the essential catalytic properties of the enzyme being changed, and that consequently the existence of multiple forms of enzymes may be expected to be a widespread phenomenon rather than an exceptional occurrence. The truth of these assumptions has been demonstrated repeatedly in recent years, so that the existence of enzymes in multiple forms has become accepted as an important generalisation in biochemistry.

Growing recognition of the widespread occurrence of multiple forms of enzymes in the later 1950's led Markert and Møller to propose the term *isozymes* to describe this type of molecular variation.[1] This term has been widely adopted, although the alternative form *isoenzymes* is preferred by many authors. 'Isoenzymes' has been used in several senses by various writers. A common usage has been, and still to some extent remains, as a description of any category of enzyme heterogeneity, without any implication as to the origin of the diverse forms or the nature of the molecular differences between them. Some rather more restricted usages have also been proposed, *e.g.*, that only multiple forms of an enzyme which occur within a single species or even within a single tissue should be classed as isoenzymes, but again with the term being used solely in an operational sense. Problems have also been encountered in defining the degree of functional similarity between multiple enzyme forms that is considered necessary to bring them within a definition of isoenzymes. These problems are especially apparent when groups of enzymes of related function, such as non-specific esterases, are considered. Functional differences (*e.g.*, in substrate specificity) between such enzymes may be so marked as to suggest that they are distinct enzymes, in spite of many similarities which appear to justify their classification as isoenzymes of a single type of enzyme.

Further possibilities of refining the definition of isoenzymes can be introduced by adding criteria based on the postulated origins of multiple molecular forms of enzymes to those reflecting catalytic similarities. Genetic and structural studies have now shown that many examples of multiple forms of enzymes are determined at the level of the structural gene. Current recommendations, therefore, are that only those different molecules which exhibit a particular type of catalytic activity but which have distinct genetic origins should be classed as isoenzymes. Multiple forms of an enzyme which are not the products of distinct structural genes, or which are of unknown origin, should not be so described.[2] Enzyme variants which arise from post-genetic modifications of a single polypeptide chain are excluded from this definition of isoenzymes, whether they are generated by partial proteolysis (as in the activation of chymotrypsinogen) or by other covalent modifications, or represent different conformational or aggregational states of the polypeptide, such as those involved in allosteric regulation of enzyme activity. However, when protomers determined by different genes combine to give several distinct polymeric enzymes, the latter are regarded as isoenzymes.

Multiple forms of a given enzyme are still frequently referred to as isoenzymes, even though the information to justify their inclusion within the recommended restricted usage of this term may be lacking. Furthermore, the experimental methods of separation and characterisation described in this monograph can be applied equally to multiple enzyme forms arising from non-genetic causes and to genetically determined isoenzymes. However, the extent to which the results of these methods can be used to deduce the probable origin of the observed enzyme heterogeneity varies from one method to another. A genetic cause of enzyme variation can often be inferred from studies of inheritance of the multiple forms, but a definitive proof of distinct genetic origins requires the demonstration of differences in primary structures (amino-acid sequences) between isoenzymes. However, the isoenzyme systems for which this level of knowledge has been achieved are relatively few. This monograph is concerned with methods which are widely used for the recognition and characterisation of multiple forms of enzymes; although these methods fall short of providing definitive structural information, it is upon them that the greater part of current knowledge of enzyme variation is based.

Genetic Causes of Enzyme Variants (Isoenzymes)

Multiple structural genes which determine the amino-acid sequences of functionally similar enzymes may have come into existence in various ways. An ancestral gene determining the structure of a particular enzyme may have undergone duplication during the course of evolution, giving rise to two distinct gene loci. These loci may subsequently have undergone separate mutations, so that the proteins which they determine have diverged in their structures, while retaining similar catalytic abilities. Alternatively, originally distinct genes and their product enzymes may have converged through successive mutations, with the result that the respective enzymes have developed similar catalytic functions. The various classes of proteases, those containing a serine residue at the active centre, those with an active sulphydryl group, and the metal-dependent proteases, may have originated in this way. A greater degree of structural homology would be expected between isoenzymes originating from divergent evolution than between those resulting from convergent evolution. For example, considerable similarities of primary structure are found between the isoenzymes of lactate dehydrogenase, which therefore have probably arisen by divergent evolution. Several enzymes exist in mitochondrial and cytoplasmic forms with markedly distinct properties and this category of isoenzymic variation may be due to convergent evolution. However, even here, considerable structural similarities may exist, as has been shown for the mitochondrial and cytoplasmic aspartate aminotransferases (EC 2.6.1.1).

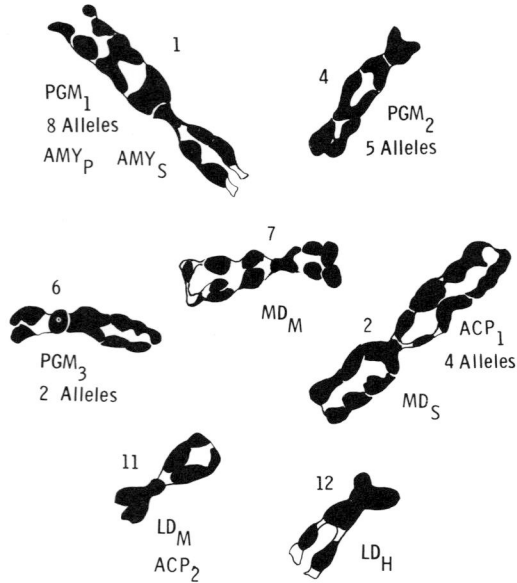

Figure 1.1 *Assignment of genes determining the structures of some isoenzymes to various human chromosomes. The genes determining isoenzymes which are structurally and functionally very similar are not necessarily closely linked on the chromosomes. The numbers of common allelic variants at certain loci are indicated; although allelic variation is not unknown at other loci, it is infrequent. (Abbreviations: ACP_1 and ACP_2, red-cell and tissue acid phosphatases; AMY_P and AMY_S, pancreatic and salivary amylases; LD_H and LD_M, 'heart' and 'muscle' lactate dehydrogneases; MD_S and MD_M, cytoplasmic and mitochondrial malate dehydrogenases; PGM_1, PGM_2, and PGM_3, phosphoglucomutases)*

Whatever their evolutionary origins, a substantial proportion of enzymes seem to be determined by more than one structural gene locus and therefore exist in isoenzymic forms. A survey of 66 different human enzymes showed no fewer than 24 to be the products of more than one gene locus.[3] Multiple gene loci have become disseminated throughout the whole species during the course of evolution, so that these genes and their dependent isoenzymes typically are present in all individuals. The several genes which determine a particular group of isoenzymes are not necessarily closely linked on the chromosomes; indeed, they may be located on different chromosomes (Fig. 1.1).

A large number of enzymes of human origin exist in multiple molecular forms which differ in type or distribution from one individual to another. Family studies show that these specific isoenzyme patterns are inherited according to Mendelian laws. They originate from modified genes, or alleles, at various chromosomal loci (Fig. 1.1). The proportion of loci subject to allelic variation has been estimated to be as high as 28%, based on an electrophoretic study of the enzymes determined by 71 human gene loci.[4] The probability that individual human beings will differ to some degree in their isoenzyme patterns is correspondingly high, and is further increased by the possibility that individuals may be heterozygous for particular mutant alleles, *i.e.*, that unlike pairs of alleles occur at one or more gene loci.

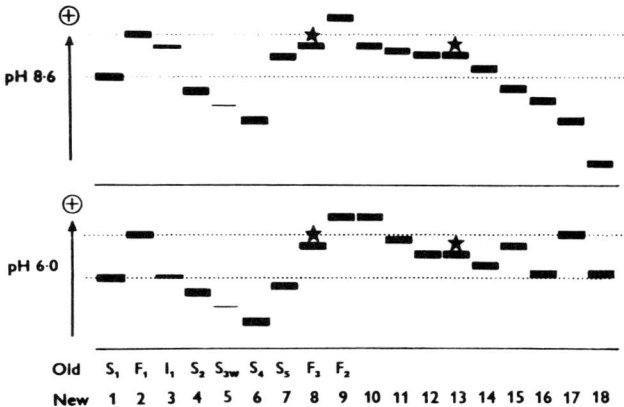

Figure 1.2   *Main zones of alkaline phosphatase activity in extracts of human placentae separated by horizontal starch-gel electrophoresis at two pH values. The electrophoretically distinct forms are determined by 18 allelic genes at a single locus. Variants marked with an asterisk cannot be distinguished solely by electrophoresis of the native isoenzymes.* [Reproduced, with permission, from: L. J. Donald and E. B. Robson, *Ann. Hum. Genet.* (Cambridge University Press), 1974, **37**, 303]

The number of allelic variants and the frequency with which particular variants occur within the population vary considerably from one enzyme to another. Mutations at either of the two principal loci which determine human lactate dehydrogenase (EC 1.1.1.27) are extremely rare, but there are a high incidence and wide variety of mutant alleles at the single locus which determines the structure of placental alkaline phosphatase, EC 3.1.3.1 (Fig. 1.2).[5] More than 150 distinct forms of the glucose-6-phosphate dehydrogenase (EC 1.1.1.49) of human erythrocytes have now been identified, together with a further 50 or so awaiting verification, each determined by a different allele at the locus on the X-chromosome which codes for this enzyme. Some of these alleles are extremely rare, whereas others occur with appreciable frequencies in particular populations or geographical locations.

Many, perhaps most, variant alleles are due to 'point mutations'; that is, the substitution of one purine or pyrimidine base for another in the triplet which codes for a particular amino acid in the polypeptide chain, so that a different amino acid is substituted. However, because of the high degree of redundancy in the genetic code, in which several different triplets code for each amino acid, only about three-quarters of such mutations will result in a change in amino acid composition. A small proportion of point mutations will cause premature termination of the growing polypeptide chain and consequent deletion of a sequence of amino acids normally present. Mutation may also result in chain-elongation. The effects of mutation on the properties of the isoenzymes determined by the mutant alleles vary from no detectable change at one extreme, to a total loss of catalytic function at the other.

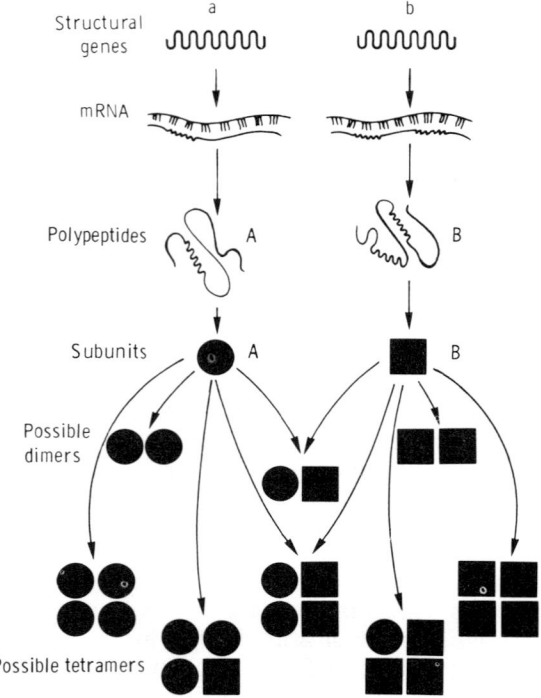

Figure 1.3 *Diagrammatic representation of the origin of dimeric or tetrameric isoenzymes by association of unlike polypeptide subunits* A *and* B, *determined by structural genes* a *and* b

When the structure of a particular enzyme is determined by multiple gene loci and when one or more of these loci are themselves subject to allelic variation, complex mixtures of isoenzymes are produced. These mixtures can be separated into patterns of multiple isoenzyme zones by electrophoresis when the component isoenzymes differ in net molecular charge. Characteristic isoenzyme patterns are associated with the occurrence of particular alleles and are further complicated in individuals who are heterozygous (*i.e.*, in whom unlike alleles are paired) at one or more loci. If the enzyme concerned consists of only a single polypeptide chain the isoenzyme pattern of heterozygotes is made up of the sum of the isoenzymes determined by each of the two alleles. However, if the complete enzyme molecule consists of more than one polypeptide subunit, or protomer, the possibility exists that protomers produced by different gene loci or mutant alleles may resemble each other sufficiently closely to be capable of aggregation to form active enzyme oligomers. Thus, a further family of hybrid isoenzymes may be produced.

The number of different hybrid isoenzymes which can be formed from two non-identical protomers depends on the number of subunits in the complete enzyme molecule. For a dimeric enzyme, one mixed dimer may be added to the two dimers composed of pairs of identical subunits, while for an enzyme with four subunits three heteropolymeric isoenzymes may be formed (Fig. 1.3). Examples of hybrid isoenzymes are the mixed MB dimer of human creatine kinase (EC 2.7.3.2), and the three hybrid isoenzymes LD_2, LD_3 and LD_4 of lactate dehydrogenase. The latter have the subunit compositions H_3M, H_2M_2, and HM_3, respectively, in which H and M represent protomers produced by separate gene loci.

If the separate gene loci which determine the structures of different protomers are expressed simultaneously in a particular tissue, the isoenzyme composition of that tissue will consist of both homopolymeric and heteropolymeric isoenzymes, although the relative proportions of the several forms will depend on the relative abundance of the different protomers. Certain gene loci are not equally active in all tissues, so that the protomers encoded by inactive loci may not be available for the formation of hybrid isoenzymes *in vivo*. For example, a third locus which determines the structure of a specific lactate-dehydrogenase subunit is active in mature human testis, and homotetramers of this subunit account for a sixth isoenzyme of lactate dehydrogenase observed in this tissue. However, this locus and the two main lactate-dehydrogenase loci are not simultaneously active in testis, or in other tissues, so that hybrid isoenzymes involving the protomer determined by the third locus are not found in tissue extracts, although hybrid isoenzymes formed from the testicular and non-testicular protomers can be produced by hybridisation experiments *in vitro*.

Since protomers determined by multiple alleles may also be capable of forming hybrid isoenzymes *in vivo*, the isoenzyme patterns of heterozygous individuals may contain isoenzymes in addition to the sum of those present in homozygotes for each of the two alleles.

Non-genetic Causes of Enzyme Heterogeneity

Processes which may operate to produce multiple forms of enzymes after the polypeptide chains have been synthesised are potentially numerous, but it is not clear to what extent, if at all, some of these processes are responsible for observed enzyme heterogeneity. Multiple forms of enzymes resulting from post-synthetic modifications (referred to as 'secondary isoenzymes' by some authors) may be generated *in vivo*, or may be artefacts associated with certain extraction procedures or particular storage conditions. Post-genetic modifications taking place in living tissue may be as significant in the understanding of certain biological processes, or as useful in clinical diagnosis, as isoenzymic variation arising at the level of the structural gene. Enzyme modification *in vitro* is also not without interest, since it can contribute to the elucidation of enzyme structure and assist in differentiating between similar forms of an enzyme. Furthermore, recognition of possible artefactual modifications may be essential for the correct interpretation of isoenzyme patterns.

Modification of the covalent structure of the polypeptide chains of certain enzymes may take place within the tissues, with resultant changes in physicochemical properties. Five forms of rabbit-muscle aldolase (EC 4.1.2.13) can be separated by isoelectric focusing. These are due to the intracellular conversion of a single asparagine residue near the carboxyl terminal of the polypeptide chain into an aspartyl residue, causing an increase in net negative charge.[6] Since the active form of this enzyme is a tetramer, hybrid molecules of native and modified chains account for the observed heterogeneity of the crystalline enzyme. Deamidation has also been identified as the cause of some of the molecular heterogeneity exhibited by certain other enzymes, such as carbonic anhydrase and amylase. Polypeptide chains may also be partially degraded by proteolytic enzymes, either within their parent tissues or their secretions, or during extraction, generating multiple enzyme forms. The family of chymotrypsins produced during partial proteolysis of chymotrypsinogen under various conditions has already been mentioned, and additional variant forms of hexokinase from yeast have been shown to be due to the action of proteases on two distinct types of enzyme molecule during extraction from the cells.[7]

Disruption of cells for the extraction of enzymes may present opportunities for interaction between tissue constituents and enzymes which do not exist in the intact cells, causing enzyme modification. Many erythrocyte enzymes, including adenosine deaminase, acid phosphatase, and some forms of phosphoglucomutase, contain sulphydryl groups which are susceptible to oxidation and in haemolysates this may be brought about by the action of oxidised glutathione, although in the intact cells this compound is present in its reduced form. Thus, variant enzyme molecules with altered molecular charge may be generated.

Covalent modification of ionisable groups present in certain amino acid side-chains of proteins can also produce enzyme forms with altered molecular charge and therefore with distinctive electrophoretic mobilities. An example of this is acetylation of the terminal amino groups of lysine and arginine, which reduces the basicity of these groups and therefore increases the anodal mobility of the protein molecule. Acetylating systems are present in some tissues, but it is not certain that acetylation of protein side-chains takes place *in vivo*. However, acetylation and other covalent modifications have been used to prepare electrophoretically altered forms of enzymes such as alkaline phosphatase *in vitro*.[8]

Modifications affecting non-protein components of enzyme molecules may also lead to molecular heterogeneity. Many enzymes are glycoproteins, and variations in carbohydrate side-chains are a common cause of non-homogeneity of preparations of these enzymes. Some carbohydrate moieties, notably *N*-acetyl neuraminic acid, are strongly ionised and consequently have a profound effect on some properties of enzyme molecules in which they occur. For example, removal of terminal sialic acid groups from human kidney alkaline phosphatase with neuraminidase greatly reduces the electrophoretic heterogeneity of the enzyme,[9] and this treatment also modifies considerably the solubilities of other human alkaline phosphatases in ethanol.[10] The addition of carbohydrate residues to the side-chains of glycoproteins appears to be a function of specific glycosyl-transferring enzymes; therefore, the possibility exists of genetic control, through these enzymes, of the structures of the carbohydrate components of glycoproteins. Genetic control of this type has been demonstrated for the water-soluble, blood group-specific glycoproteins. Glycosyltransferases determined by A or B alleles at the ABO gene locus catalyse the addition of either *N*-acetyl-D-galactosamine residues or D-galactose residues, respectively, to the terminal positions of the polysaccharide chains of these glycoproteins. Structural variations in the carbohydrate components of multiple forms of enzymes may similarly be manifestations of the differential expression of multiple gene loci or alleles which control glycosyl transferases. However, since structural variants arising in this way would not involve the primary structures of the multiple enzyme forms (*i.e.*, their amino-acid sequences), they would not be regarded as true isoenzymes, in spite of their genetic origins. So far, genetically determined differences in carbohydrate components have not been unequivocally identified as a cause of multiple forms of glycoprotein enzymes. Furthermore, many glycoprotein enzymes are components of structural elements of the cells in which they occur, such as cell membranes, and the rather vigorous treatments needed to bring these enzymes into solution may result in partial degradation and may contribute to the degree of heterogeneity observed in enzyme preparations.

Attachment of lipids or peptides to a protein core may be a cause of some examples of enzyme heterogeneity, *e.g.*, of mouse-intestinal alkaline phosphatase.[11] Fractionations of alcohol dehydrogenase (EC 1.1.1.1) by ion-exchange chromatography or zone electrophoresis of extracts of *Drosophila melanogaster* or horse liver are modified by the presence of the oxidised or reduced coenzyme, nicotinamide-adenine dinucleotide.[12, 13] This effect has been attributed to the alterations of molecular charge resulting from combination of the enzyme and coenzyme molecules. Interaction of enzyme molecules with anions or cations is also potentially capable of altering enzyme zones separable by charge-dependent means.

Aggregation of enzyme molecules with each other or with non-enzymic proteins may give rise to multiple molecular forms which can be separated by techniques that depend on differences in molecular size. A number of examples of enzyme heterogeneity caused in this way have been described. Multiple forms of serum cholinesterase (EC 3.1.1.8) can be separated on electrophoretic media such as starch gel in which molecular size influences segregation of the protein zones. Four catalytically active cholinesterase components with molecular weights ranging from about 80 000 to 340 000 are found in most sera, with the heaviest component, C_4, contributing most of the enzyme activity. Additional enzyme forms are also present occasionally, but it appears that the principal serum cholinesterase fractions can be attributed to different states of aggregation of a single monomer.[14] The acetyl-cholinesterase of related substrate-specificity prepared from human erythrocytes also seems capable of existing in several active polymeric forms.[15] Some components of human placental alkaline phosphatase represent aggregated forms of the enzyme,[16] while association of alkaline phosphatase with lipoproteins has been suggested as the explanation of minor active components of this enzyme migrating slowly in starch-gel electrophoresis, which are regularly present in extracts of various human tissues.[17]

A specific form of interaction between enzymic and non-enzymic proteins has been shown to be the cause of unusual enzyme components seen when some samples of human plasma are fractionated by electrophoresis or chromatography. These components are due to combination of apparently normal enzyme or isoenzyme molecules with certain plasma immunoglobulins. The enzyme–protein complexes thus formed may themselves be heterogeneous. Lactate dehydrogenase has been shown to combine with immunoglobulins A or G, and complexes between this enzyme and IgG with a range of molecular weights from 300 000 to 1 000 000 and with isoelectric points from pH 6.3 to 7.2 were found in the serum of a single patient.[18] Macroamylase, the first such enzyme complex to be identified in serum, is itself not homogeneous, with sedimentation coefficients ranging from 7.7S to above 19S.[19] Immunoglobulin A and glycogen may be involved in macroamylase formation, and the complexes appear to have a greater affinity for salivary amylase than for the pancreatic isoenzyme.[20] Alkaline phosphatase, also, has been observed as unusual zones[21] which appear to be due to complexes with IgG in serum.[22]

A single polypeptide chain can in theory exist in an infinite number of different configurations. However, one specific configuration generally appears to be the most stable for any given sequence of amino acids, and this configuration is assumed by the chain as it is synthesised within the cell. Thus, the primary structure of the polypeptide chain also determines its three-dimensional secondary and tertiary structures. It is conceivable that, in some cases, there may be several alternative configurations of a single chain which are of nearly equal stabilities and, therefore, that these alternative forms may coexist. This possibility has been invoked to explain some multiple enzyme forms.

Malate dehydrogenase (EC 1.1.1.37) exists as distinct mitochondrial and cytoplasmic isoenzymes under separate genetic control. The mitochondrial isoenzyme of malate dehydrogenase from many sources and the cytoplasmic isoenzyme from some vertebrate cells can each be separated electrophoretically into several distinct components. The zones which make up each isoenzyme appear to have nearly identical catalytic properties.[23] Malate dehydrogenase is a dimer and therefore different combinations of four non-identical subunits (two cytoplasmic and two mitochondrial) could account for up to three multiple forms of each isoenzyme. The appearance of the electrophoretic patterns of malate dehydrogenase in carriers of variant forms of the mitochondrial isoenzyme lends some support to this view.[24] However, the number of zones of mitochondrial malate dehydrogenase is greater than can be accounted for by combination of unlike monomers. It has been suggested, therefore, that conformational isomers ('conformers') may arise by small differences in folding of a single type of polypeptide chain, to give variants with identical active sites but with different net molecular charges.[25]

The original evidence in favour of conformational isomerism included the observation that incorporation of a relatively small amount of iodine into mitochondrial malate dehydrogenase isoenzymes produced significant alterations in electrophoretic mobility, accompanied by changes in optical rotatory dispersion suggestive of conformational change.[25] A

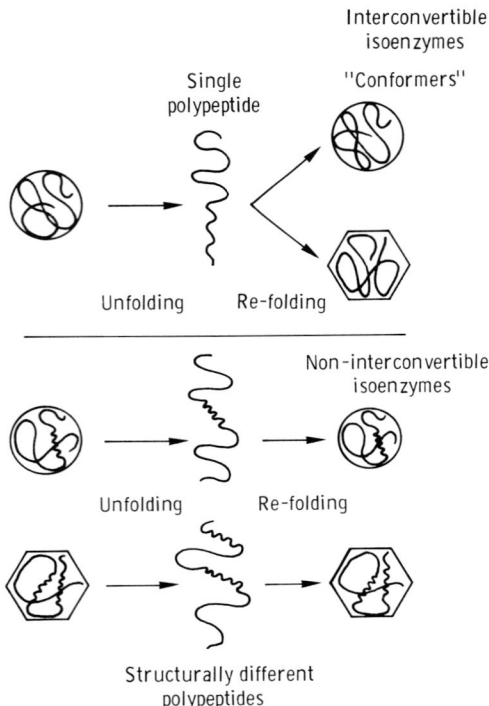

Figure 1.4   *Diagrammatic representation of the 'conformer' hypothesis. If a constituent polypeptide chain is almost equally stable in more than one conformation (above), unfolding and re-folding of a single isoenzyme will generate a mixture of the possible conformational isomers. On the other hand, chains stable in only one conformation (below) give rise to isoenzymes which are not inter-convertible by unfolding and re-folding*

prediction of the conformer hypothesis is that reversible denaturation of one of the isomers should generate the complete set, since it is implied that the several alternative conformations are of nearly equal stabilities (Fig. 1.4). However, attempts to fulfil this prediction have not been completely successful.[26] Furthermore, although reversible denaturation of individual mitochondrial isoenzymes with acid or guanidine hydrochloride did produce renatured forms with electrophoretic mobilities similar to those of native isoenzymes, the native and renatured forms were not identical in stability to heat.[27] Conformational isomerism has been proposed as an explanation for other examples of enzyme heterogeneity besides the multiple forms of mitochondrial malate dehydrogenase, *e.g.*, of aspartate aminotransferase, the MM-isoenzyme of creatine kinase, and erythrocyte acid phosphatase, but in no case has this cause of multiple enzyme forms been established unequivocally.

Conformational changes affecting subunits of enzyme molecules have also been invoked to account for the co-operative effects of increasing substrate concentration characteristic of some enzymes, in which binding of one molecule of substrate facilitates the subsequent attachment of substrate molecules at other active centres. Conformational changes have been shown by crystallographic techniques to accompany binding of substrate molecules to some enzymes and the attachment of oxygen molecules to haemoglobin. Enzymes with co-operative substrate-binding characteristics are termed 'allosteric' enzymes, and their binding properties are often found to be modified further by activators, which facilitate substrate binding, or inhibitors, which reduce it. These modifiers are also thought to act by inducing conformational changes when they are bound to specific modifier-binding sites. Like the postulated 'conformational isoenzymes', the different conformational states of an allosteric enzyme induced by combination with substrates or modifiers are not regarded as isoenzymes, since no changes in primary structure of the enzyme molecule are involved. Nevertheless, some of the techniques of isoenzyme characterisation described later, such as selective inactivation, may be able to distinguish between certain alternative enzyme conformations.

Distribution of Multiple Forms of Enzymes in Tissues

The existence of multiple gene loci and the isoenzymic adaptations which this makes possible have presumably conferred an evolutionary advantage on the species and have thus become part of its normal metabolic pattern. Some of these adaptations are obviously related to the division of function between and within different types of specialised cells and tissues which has accompanied the evolution of higher organisms. It is to be expected, therefore, that the distribution of isoenzymes will not be uniform throughout the organism and wide variations in the activity of different isoenzymes do indeed occur in man as in other species between organs, between the cells which compose a particular organ, and even between the structures which constitute a single cell. Tissue-specific differences in distributions of other multiple forms of enzymes are also found, some of which cannot be attributed with certainty to the existence of multiple gene loci. Whatever their molecular origins, however, well-established differences in distribution account for much of the value of studies of multiple enzyme forms in medicine, not only by providing insights into the metabolic patterns of different tissues and their derangements in disease, but also as the basis for organ-specific diagnosis by isoenzyme measurements.

When several gene loci determine a particular type of enzyme activity, the respective isoenzymes determined by the different loci are present in each tissue in which the loci are active, together with any hybrid isoenzymes which may be formed between the different gene products. Homologous isoenzymes prepared from different tissues are identical, but the relative proportions of the isoenzymes in a given tissue depend on their biological half-lives; *i.e.*, on the resultant effects of differential rates of formation and breakdown of the various isoenzymes.

The total lactate dehydrogenase activity of most human and animal tissues is contributed by five isoenzymes. The relative proportions of the isoenzymes differ from tissue to tissue, but three basic patterns can be distinguished (Fig. 1.5). In tissues such as cardiac muscle, kidney, and erythrocytes the electrophoretically faster, *i.e.*, more-anodal, isoenzymes LD_1 and LD_2 predominate, whereas in liver and skeletal muscle the more-cathodal LD_4 and LD_5 isoenzymes are prominent. Isoenzymes of intermediate mobility account for the lactate dehydrogenase activity of

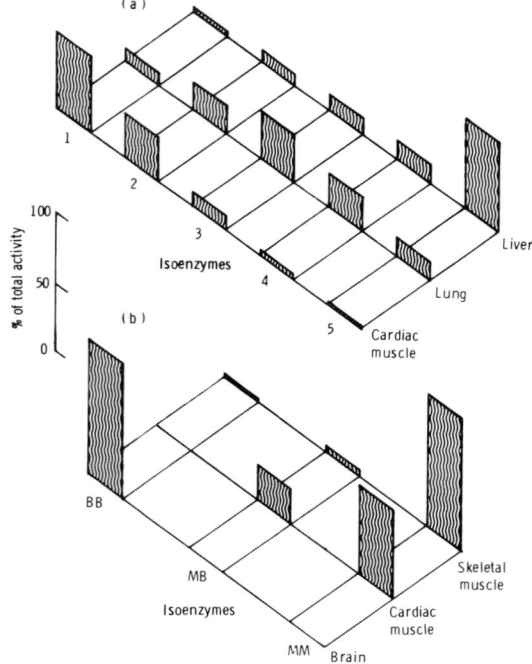

Figure 1.5 *Distribution of enzyme activity among the individual isoenzymes of (a) lactate dehydrogenase, and (b) creatine kinase in some human tissues. In each case the electrophoretically more-anodal forms are at the left*

many tissues, *e.g.* endocrine glands, spleen, and lymph nodes, and non-gravid uterine muscle. A different, sixth lactate dehydrogenase isoenzyme, LD_X, is present in post-pubertal human testis.

The distribution of lactate dehydrogenase isoenzymes within tissues has been less thoroughly studied, but some variations in isoenzyme content between different regions or cells of certain tissues have been reported. A higher proportion of LD_1 has been found in the cortical regions of rat kidney than in the medulla, although human kidney cortex and medulla appear to differ little in isoenzyme composition.[28] LD_5 was almost entirely responsible for the lactate dehydrogenase activity of purified parenchymal cells from rat liver, whereas the electrophoretically faster isoenzymes LD_4, LD_3, and LD_2 were also present in significant amounts in Küpffer cells.[29] Particular interest attaches to the distribution of lactate dehydrogenase isoenzymes amongst muscle fibres of different types because of the postulated differences in function between the isoenzymes. In species in which red (or Type I) muscle fibres can be distinguished clearly from white (or Type II) fibres, the latter contain chiefly LD_5 and the former a greater proportion of the more-anodal isoenzymes. A distinction between red and white skeletal muscles cannot be drawn readily in man, most human muscles containing a mixture of fibre types. However, such studies on separated fibre types from human muscle as have been made suggest that the distribution of lactate dehydrogenase isoenzymes in human muscle is broadly similar to that in other vertebrates. Red fibres obtained from human skeletal muscle by microdissection more frequently showed the faster lactate dehydrogenase isoenzymes and white fibres the slower forms.[30] The faster isoenzymes were also more prominent in muscles in which histochemical methods indicated the presence of a high proportion of Type I fibres, but this correlation between lactate dehydrogenase isoenzyme distribution and apparent type of fibre was not constant.[31]

The relative proportions of the H and M subunits available to make up active lactate dehydrogenase tetramers (see p. 7) are determined by the rates of both formation and breakdown of the two polypeptides in a particular tissue. In the rat, LD_5 (the M_4 tetramer) is synthesised about four times faster in liver than in heart and about twice as fast in skeletal muscle. This isoenzyme is also degraded 10 to 20 times more rapidly in heart, so that the half-life of LD_5 in heart is a tenth or less of that in the other two tissues.[32] This implies that, in tissues in which the half-life of one gene product greatly exceeds that of the other, almost all of the lactate dehydrogenase activity will exist in the form of tetramers of that polypeptide, *i.e.*, as either mainly LD_1 (H_4) or LD_5 (M_4). The smaller amounts of the less abundant subunit will be present as decreasing amounts of the mixed tetramers, *e.g.*, H_3M, H_2M_2, and HM_3 in tissues in which H-subunits are predominant. The homopolymer of the less favoured polypeptide (*e.g.*, M_4) may be undetectable. When the synthetic activity of the two genes is nearly equal and their products have similar half-lives the most common isoenzyme is LD_3, the H_2M_2 tetramer, since association of the subunits into active tetramers is random.

The existence and differential activities of two structural genes, each determining the amino-acid sequence of a specific polypeptide, also accounts for multiple forms of creatine kinase and their specific distributions in human and animal tissues (Fig. 1.5). However, the active form of this enzyme is a dimer, so that only three different pairs of subunits can be formed to produce active isoenzymes. The electrophoretically most-anodal homopolymer is present in large amounts in human brain, and it is consequently designated BB and its constituents as B subunits. The BB dimer is also the form of creatine kinase present in prostate, thyroid, kidney, stomach, bladder, and lung, but in smaller amounts than in brain. The more-cathodal MM dimer, made up of M subunits, is the predominant enzyme form in both cardiac and skeletal muscle. There is general agreement that cardiac muscle also contains a substantial proportion of the isoenzyme form with intermediate electrophoretic mobility and which consists of MB dimers, variously estimated as being between 10 and 50% of the total activity of the tissue. This range of estimates is partly accounted for by inter-species differences in isoenzyme distribution; *e.g.*, the contribution of the MB isoenzyme to the total creatine kinase activity is considerably lower in canine than in human heart. The proportion of MB isoenzyme in human skeletal muscle is considerably less than in cardiac muscle. This isoenzyme may be undetectable by some procedures in extracts of skeletal muscle, and accounts for much less than 10% of the creatine kinase activity of this tissue. The MM isoenzyme of creatine kinase may itself be heterogeneous. Evidence has been obtained from isoelectric focusing experiments that two distinct M subunits may exist, each capable of forming active dimers with other M or B subunits.[33]

Tissue-specific distributions of several other human enzymes have been shown to result from the formation of homo- and hetero-polymers between polypeptide subunits which are themselves determined by separate gene loci. These enzymes thus provide further examples of the type of heterogeneity displayed by lactate dehydrogenase and creatine kinase. Alcohol dehydrogenase exists in human tissues in multiple forms which vary in their relative activities in different tissues. The active enzyme molecule is a dimer, and three gene loci appear to control the structures of three different subunits. Heteropolymers and homopolymers can form by association of any two of these subunits, to give three-membered sets of isoenzymes *in vivo* in tissues in which two loci are active. The dissociation and recombination of subunits can also be reproduced *in vitro*.

Aldolase resembles lactate dehydrogenase in that it is a tetrameric enzyme with subunits determined by three separate loci. As with lactate dehydrogenase, only two of the loci, those producing A and B subunits, appear to be active simultaneously in most tissues, so that the most common isoenzyme pattern is of varying proportions of the components of a five-membered isoenzyme set of which two members correspond to the A and B homopolymers. The locus which determines the structure of the C subunit is active in brain tissue, as is the A locus, so that this tissue contains aldolases A and C together with the three corresponding heteropolymers.

Certain gene loci may be expressed exclusively, or almost so, in a single tissue only; perhaps at a particular stage in the development of the organism. An example which has already been mentioned is the third lactate-dehydrogenase locus, active in mature testis, which determines the third homopolymer of the enzyme, LD_X. The isoenzyme of alkaline phosphatase which occurs in the human placenta is the product of a single structural gene locus, distinct from loci which specify the structures of other forms of alkaline phosphatase, and the product of the placental-phosphatase locus is normally detectable only in the placenta.

Alkaline phosphatases occur in many other human and animal tissues and some of these forms of the enzyme have distinctive properties, such as characteristic electrophoretic mobilities (Fig. 1.6). Present evidence

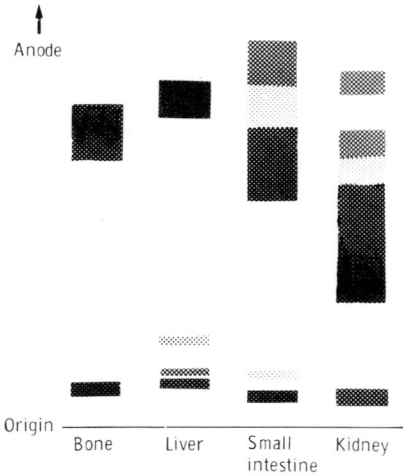

Figure 1.6 *Distribution and relative intensities of zones of alkaline phosphatase activity in extracts of some human tissues separated by starch-gel electrophoresis. In contrast to the distributions of lactate dehydrogenase and creatine kinase isoenzymes (Fig. 1.5), no single zone of alkaline phosphatase is identical in mobility in all four tissues*

does not allow a definite statement to be made as to the number of structural gene loci which determine the various non-placental alkaline phosphatases, except that the enzyme from small intestine is almost certainly the product of an unique locus.[34] However, in spite of the considerable electrophoretic heterogeneity of alkaline phosphatase within each tissue, which probably results from the formation of various complexes containing the enzyme as well as from variations in carbohydrate components,[9, 11, 17, 35] the total alkaline phosphatase activity of a particular tissue seems essentially to be due to the presence in it of a single form of the enzyme. Thus, if separate structural gene loci do determine the structures of non-placental, non-intestinal alkaline phosphatases such as those of bone, liver, and kidney, these loci are not expressed simultaneously in different tissues.

Exceptions to this generalisation can be found in some human tissues, in which minor alkaline phosphatase components resembling the forms predominant in other organs can be found. Thus, extracts of kidney contain minor components of alkaline phosphatase which are unaffected in net molecular charge by treatment with neuraminidase,[9] a property possessed by intestinal phosphatase, and which cross-react with antiserum to intestinal phosphatase.[36] Conversely, a small fraction of intestinal alkaline phosphatase is neuraminidase-sensitive.[37] Different cell types which together make up a single organ may each contribute a specific alkaline phosphatase variant to the total alkaline phosphatase activity. There is some evidence that, when parenchymal cells and biliary-tract cells of rat liver are separated from each other by perfusion of the organ with collagenase solution, the alkaline phosphatase extracted from the parenchymal cells has a greater anodal mobility on electrophoresis and is more stable to heat than the enzyme from non-parenchymal cells.[38] However, it is not certain whether the non-parenchymal and parenchymal alkaline phosphatases are distinct isoenzymes, or whether they arise by cell-specific, post-genetic modifications.

A particularly striking example of the localised expression of multiple gene loci is provided by distinct isoenzymes which occur exclusively in specific subcellular organelles. Differences between mitochondrial isoenzymes and their functionally analogous counterparts in the cytoplasm have been demonstrated in several cases. That the mitochondrial and extra-mitochondrial enzyme forms are indeed determined by separate gene loci, *i.e.*, that they are true isoenzymes, has been confirmed for human aspartate aminotransferases and malate dehydrogenases from these two subcellular locations by the discovery of rare variants of the respective isoenzymes from mitochondria or cytoplasm, which are inherited in a Mendelian manner without corresponding changes in the isoenzymes located elsewhere in the cell.

The mitochondrial isoenzymes of aspartate aminotransferase and malate dehydrogenase account for about one-half and two-thirds, respectively, of these two activities in the whole cell. Although isoenzymes of non-mitochondrial organelles have been less well characterised than the enzyme variants found in mitochondria, differences in properties do exist between catalytically similar enzymes prepared from various other subcellular fractions, *e.g.*, cytosol and plasma membranes. It is probable that some of these variations also arise from the presence of multiple structural gene loci.

Developmental Changes in Isoenzyme Distribution.—Multiple gene loci and their dependent isoenzymes provide means for the adaptation of metabolic patterns to the changing needs of different organs and tissues in the course of normal development, or in response to environmental change. Pathological changes, also, may be associated with alterations in the activities of specific isoenzymes. Both physiological and pathological alterations in isoenzyme patterns are important, not only for the insights they provide into normal and abnormal metabolism, but also in the use of isoenzyme studies in diagnosis.

(*a*) *Changes during normal development.* The patterns of several sets of isoenzymes change during normal development in tissues from many species. Although the nature and time-course of the changes often differ between species, especially between lower and higher animals, the distribution of isoenzymes at a particular moment in ontogeny may be assumed to reflect the needs of the developing tissues.

Changes in the relative proportions of several isoenzymes are seen during the embryonic development of skeletal muscle.[39, 40] The proportions of the electrophoretically more-cathodal isoenzymes of both lactate dehydrogenase and creatine kinase increase in this tissue, so that the qualitative patterns associated with the differentiated muscle are present by about the sixth month of intra-uterine life. Smaller, quantitative changes in isoenzyme distribution may continue to birth and into early post-natal life. An increased proportion of anodal lactate dehydrogenase isoenzymes has been noted in muscle tissue of normal subjects over 60 years of age, resembling the pattern seen in young children.[31]

As might be expected from the highly specialised nature of the metabolism of the liver, this tissue also shows characteristic changes in the patterns of several isoenzymes during embryogenesis. In early foetal development, all three aldolase isoenzymes, A, B, and C, together with the various hybrid tetramers, can be detected in extracts of liver.[41] However, at birth aldolase B is the predominant isoenzyme, as in adult liver. Striking changes in the distribution of isoenzymes of alcohol dehydrogenase occur in human liver during pre-natal development. A single alcohol dehydrogenase isoenzyme is present in embryonic liver after about 10-weeks gestation, consisting of dimers of the α-polypeptide controlled by the ADH_1 locus. Increasing activity of the second ADH locus is shown by the appearance of β-polypeptides, seen first as hybrid dimers with α-polypeptides and then as $\beta\beta$-dimers, so that, at birth, three alcohol dehydrogenase isoenzymes, ($\alpha\alpha$, $\alpha\beta$, and $\beta\beta$) are present.

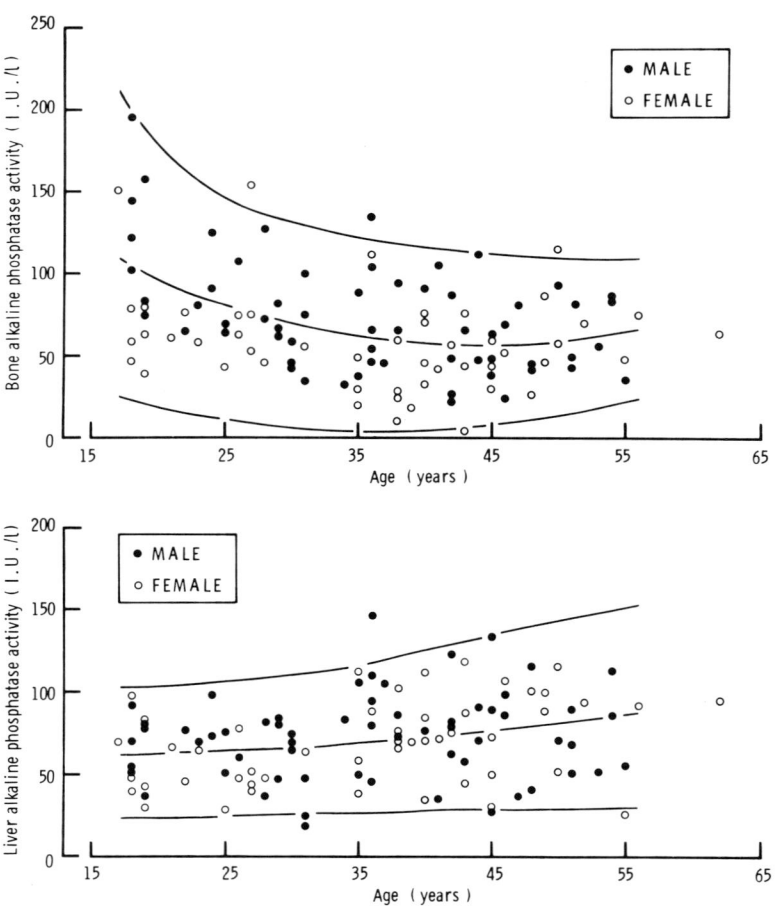

Figure 1.7   *Changes in the activities of bone (above) and liver (below) forms of alkaline phosphatase with age in the serum of normal subjects.* [Reproduced, with permission, from: K. B. Whitaker, L. G. Whitby, and D. W. Moss, *Clin. Chim. Acta* (Elsevier), 1977, **80**, 209]

Continuing changes in the expression of these two loci increase the proportion of the $\beta\beta$-dimers relative to $\alpha\alpha$-dimers. Furthermore, expression of a third locus, ADH_3, leads to the appearance of γ-polypeptides and additional dimeric isoenzymes.[42]

The changes in isoenzyme patterns during development discussed so far are presumed to result from changes in the relative activities of multiple gene loci within developing cells of a particular type, *e.g.*, hepatocytes. Other alterations in the balance of isoenzymes within the whole organism may derive from changes in the number or activity of cells which contain large amounts of a characteristic isoenzyme. An example of this is the increased number and activity of the osteoblasts which are responsible for mineralisation of the skeleton between the early post-natal period and the beginning of the third decade of life. These cells progress through a life cycle in which the activity of alkaline phosphatase in them increases, then falls again as the cells become surrounded by bone matrix. The excess of alkaline phosphatase from the active osteoblasts enters the circulation, where its presence can be recognised by its characteristic properties and where it elevates the total serum alkaline phosphatase activity of young subjects above that of skeletally mature adults (Fig. 1.7). An alkaline phosphatase isoenzyme apparently originating in the liver also contributes to the total activity of this enzyme in normal plasma, and the amount of this isoenzyme in plasma shows a small, progressive increase with age[43] (Fig. 1.7). The reason for this age-dependent change is not known, but it may result from increased synthesis of the isoenzyme by hepatocytes in response to continuing exposure to factors which induce enzyme synthesis.

Changes have been observed in the characteristics of certain isoenzymes as their parent cells become older; *e.g.*, of glucose-6-phosphate dehydrogenase in erythrocytes and in ageing human fibroblasts in cell cultures.[44,45] These and similar alterations in isoenzyme properties may reflect the accumulation of errors in protein biosynthesis during ageing.

(b) Pathological changes in isoenzyme distribution. Certain diseases, such as the progressive muscular dystrophies, appear to involve a failure of the affected tissues to mature normally or to maintain a normal state. Cancer cells show a progressive loss of the attributes of structure and metabolism of the healthy cells from which they arise. It is to be expected, therefore, that when the mature and fully differentiated tissue displays a characteristic complement of isoenzymes, this pattern may be lost or modified if normal differentiation is arrested or reversed, and many examples of isoenzyme changes accompanying such processes have been reported.

A resemblance between dystrophic human skeletal muscle and foetal muscle in their isoenzyme patterns has been demonstrated in many studies, though not in all. The distributions of isoenzymes of aldolase, lactate dehydrogenase and creatine kinase in the muscles of patients with progressive muscular dystrophy have been found to be similar to those seen in the earlier stages of development of foetal muscle.[39,46,47] The isoenzyme abnormalities seen in dystrophic muscle have been interpreted as a failure to maintain, or even to reach, a normal degree of differentiation. The rather similar, though less pronounced, changes which have been observed in ageing or disused muscle may also represent de-differentiation, with alterations in the function of regulatory mechanisms which normally determine the relative proportions of the various isoenzymes in the differentiated tissue. Isoenzyme patterns may also show some tendency to approach foetal distributions in regenerating tissues. This may result from relaxation or modification of control systems in rapidly dividing cells and may account for some of the isoenzyme changes seen, for example, in muscle in acute polymyositis.

A re-emergence of foetal patterns of isoenzyme distribution is also a feature of malignant transformation in many tissues. This phenomenon was first studied extensively in the case of lactate dehydrogenase isoenzymes. Malignant tumours in general show a significant shift in the balance of isoenzymes towards the electrophoretically more-cathodal forms, LD_4 and LD_5.[48] The decline in activity of the LD_1 and LD_2 isoenzymes results in patterns which are reminiscent of those occurring in embryonic tissues. Tumours of prostate, cervix, breast, brain, stomach, colon, rectum, bronchus, and lymph nodes are among those which show this transformation. In contrast, comparatively benign gliomas show a relative increase in anionic isoenzymes. A relative increase in the proportion of cathodal isoenzymes of lactate dehydrogenase has also been observed in tissues adjacent to malignant tumours, *e.g.*, of the colon, although the cells in these regions are morphologically normal.[49] These changes may represent an early manifestation of metabolic alteration in cells which subsequently become malignant.

As well as lactate dehydrogenase, the isoenzyme patterns of aldolase, pyruvate kinase (EC 2.7.1.40), and hexosaminidase (EC 3.2.1.30) have been shown to undergo a change towards foetal-like patterns in hepatoma.[50] These changes were also present, but to a less marked extent, in regenerating liver. A further eleven isoenzyme systems which differ in neoplastic cells from their counterparts in normal tissues have been listed by Criss, including examples such as malate dehydrogenase and aspartate aminotransferase in which the balance between cytoplasmic and mitochondrial isoenzymes is altered.[51]

In 1968, Fishman and his collaborators reported the identification of an alkaline phosphatase in the serum of a patient with metastatic squamous cell carcinoma of the lung which was identical with the alkaline phosphatase of normal placenta with respect to inhibition by L-phenylalanine, heat-stability, and alteration of electrophoretic mobility by digestion with neuraminidase. The newly discovered isoenzyme, which was termed the Regan isoenzyme after the patient in whom it was discovered, was also precipitated by an antiserum to human placental alkaline phosphatase.[52] Further evidence of similarity between placental alkaline phosphatase and the Regan isoenzyme was provided by comparison of the purified placental isoenzyme with a preparation of alkaline phosphatase from liver metastases of a giant-cell carcinoma of the lung. The two enzymes had identical subunit molecular weights and their isoelectric points were closely similar. The placental and tumour-derived alkaline phosphatases had identical N-terminal amino-acid sequences over the last four residues, and this sequence was different from that of the liver isoenzyme. Two-dimensional peptide maps of partially digested alkaline phosphatases of tumour and placenta showed slight differences, but these differences were close to the limits of experimental variation. These results supported the view that these two isoenzymes were products of the same structural gene with some small differences resulting from post-genetic modification.[53] Regan isoenzymes from different sources display a range of electrophoretic mobilities, analogous to the different mobilities of the normal phenotypes of placental alkaline phosphatase.

The Regan isoenzyme originally described accounted for about half of the elevated alkaline phosphatase activity of the serum and it was present in large quantities in extracts of the tumour tissue. Since its first description, the Regan isoenzyme has been detected in the sera of patients with many types of malignant disease, and also in some patients with non-malignant diseases. An incidence of the isoenzyme of 3 to 15% in cancer patients has been estimated, but this varies with the sensitivity of the methods used for its detection. Other variant forms of alkaline phosphatase have since been discovered in tumour tissues. These variants also show many similarities to both normal placental alkaline phosphatase and the Regan isoenzyme, but differ from them in such properties as response to some inhibitors.

Reversions to more primitive patterns of isoenzyme distribution are not uncommon in cells cultured *in vitro*. These changes and those described in neoplastic tissues appear to be due to a general loss of differentiation, and de-repression of genes which are normally only expressed during embryogenesis. Somatic mutations or post-genetic modifications may cause the isoenzymes appearing in cancer cells to differ somewhat from their counterparts in normal cells. Attempts have been made to link changes in isoenzyme distribution in cancer cells with altered metabolic patterns, *e.g.*, increased glycolysis. Although such correlations can plausibly be made for some isoenzymes, notably those of lactate dehydrogenase, not all altered distributions of isoenzymes in cancer cells can be attributed to adaptations to specific patterns of metabolism.

Many examples are known of induction of enzyme activity, by which exposure to certain agents such as drugs increases the activity of a particular enzyme by stimulating increased synthesis of it. When an inducing agent acts selectively on a particular isoenzyme, the balance of isoenzymes making up the total enzymic activity in question is changed. For example, several different kinds of stimuli have been shown selectively to increase the alkaline phosphatase activity of various tissues. These include dietary phosphate, which influences rat kidney alkaline phosphatase, vitamin D, which induces intestinal alkaline phosphatase in chicks, and cortisone or similar substances, which produce an increased activity of alkaline phosphatase in human leukocytes, mouse duodenum and in some cultured HeLa cell lines. Catecholamines or their synthetic analogue isoprenaline cause an increased alkaline phosphatase activity localised specifically to the right atrium of rat heart. However, one of the most interesting of such phenomena from the clinical point of view is the increase in alkaline phosphatase activity in the liver which follows occlusion of the bile duct. For many years, the increased activity of alkaline phosphatase in serum seen in pathological or experimental obstruction of the bile duct was attributed to retention of alkaline phosphatases from extra-hepatic sources, chiefly bone, which, it was presumed, would normally be excreted in the bile. However, characterisation of alkaline phosphatases from serum and tissues revealed that the enzyme in serum in cholestasis resembled liver alkaline phosphatase kinetically and electrophoretically rather than the isoenzyme from bone.[54,55] Direct evidence that the liver was the source of increased alkaline phosphatase activity in biliary obstruction was provided subsequently by experimental occlusion of the bile duct from a single lobe of dog liver or from isolated, perfused cat liver, resulting in increased activities of the enzyme in the hepatic tissue.

The rise in alkaline phosphatase activity in the majority of the foregoing examples is most probably due to increased synthesis of a pre-existing alkaline phosphatase isoenzyme. This is supported by the general similarity of properties of the basal and induced alkaline phosphatases, and by the effect of inhibitors of protein biosynthesis which generally prevent the increase of enzyme activity. A similar explanation appears to account for most instances of induced increases in other enzyme activities. However, alternative possibilities include an increase in catalytic efficiency of pre-existing enzyme molecules. This has been held to account for the enhancing effect of cortisone or its analogues on the alkaline phosphatase activity of HeLa cells, which is not accompanied by an increase in the amount of enzyme protein.[56]

An increase in the number or activity of particular cells which are rich in a specific isoenzyme will alter the balance of isoenzymes responsible for the activity of the enzyme concerned in the whole organism. Such changes have already been mentioned as an accompaniment of normal maturation, but they may also result from disease. For example, prostatic cells contain a high concentration of a specific isoenzyme of acid phosphatase (EC 3.1.3.2). When secondary deposits from a primary cancer of the prostate become disseminated to other parts of the body they usually act as additional sources of this isoenzyme.

Biological Significance of Multiple Forms of Enzymes

As already mentioned, the existence of multiple structural gene loci determining functionally similar though not identical isoenzymes provides a means by which metabolic patterns may be adapted quantitatively to the different needs of tissues while remaining qualitatively similar. This may result from the different catalytic properties of the isoenzymes, or from additional possibilities of control of enzyme activity arising from the tissue-specific expression of particular loci, or from the selective effect of activators or inhibitors on particular isoenzymes. Correlations between the properties of isoenzymes predominant in certain tissues and the metabolic patterns of those tissues have been clearly demonstrated in several cases.

Aldolase is a tetrameric enzyme with subunits determined by three separate gene loci. Only two of these loci, those producing A and B subunits, appear to be active simultaneously in most tissues, so that the most common isoenzyme pattern consists of varying proportions of the components of a five-membered set of isoenzymes, of which two members correspond to the A and B homopolymers. (The further degree of heterogeneity which is introduced into the A isoenzyme from rabbit muscle by de-amidation of an asparagine residue has already been mentioned; p. 8.) The locus which determines the structure of the C subunit is active in brain tissue, as is the A locus, so that this tissue contains aldolases A and C together with the three corresponding heteropolymers.

The properties of aldolases A, B, and C from several species have been studied extensively.[57] The catalytic differences between the A and B isoenzymes are consistent with the patterns of metabolism in the tissues in which they mainly occur. Both isoenzymes will cleave fructose-1,6-diphosphate or fructose-1-phosphate, but aldolase A shows a fifty-fold greater activity towards fructose-1,6-diphosphate than towards fructose-1-phosphate. This is in keeping with its part in the glycolytic metabolism of skeletal muscle, in which it is the predominant isoenzyme, since cleavage of fructose-1,6-diphosphate to triosephosphate is a key reaction in glycolysis. Aldolase B, the main isoenzyme of liver, shows no marked preference for the diphosphate substrate and this and other properties indicate that it is better adapted to utilisation of fructose and to gluconeogenesis. The properties of aldolase C are intermediate between those of the A and B isoenzymes, but the function of this isoenzyme is uncertain.

The H_4 (LD$_1$) and M_4 (LD$_5$) tetramers of lactate dehydrogenase also have significantly different catalytic properties, with the heteropolymers showing intermediate characteristics. The dependence of the rates of reaction catalysed by LD$_1$ or LD$_5$ on the concentrations of either lactate or pyruvate has been studied extensively in the search for clues to the physiological function of these isoenzymes. The LD$_1$ isoenzyme of lactate dehydrogenase is inhibited by excess of pyruvate to a greater extent than the LD$_5$ isoenzyme under certain conditions *in vitro*. On the basis of this observation, it has been suggested that rapid accumulation of lactate can occur in a tissue in which the main form of lactate dehydrogenase is LD$_5$, whereas if LD$_1$ is the predominant isoenzyme, its inhibition by excess of pyruvate will prevent conversion of excess of pyruvate to lactate. The ability of tissues to function anaerobically, *i.e.*, in conditions leading to accumulation of lactate, is therefore presumed to be due to the presence in them of LD$_5$.[58]

Impressive support for the hypothesis that capacity for anaerobic metabolism and the presence of LD$_5$ are correlated has been gathered from the study of the isoenzyme composition of muscles from avian species, in which a clear distinction can be drawn between muscles capable of sustained contraction and those which are active only intermittently. The former muscles which, like human cardiac muscle, operate aerobically contain the electrophoretically faster lactate dehydrogenase isoenzymes while the latter, like human skeletal muscle, contain the slower isoenzymes.[59] However, this generalisation does not appear to be universally applicable, since liver contains a large amount of LD$_5$ but has an aerobic pattern of metabolism.

The differences between the LD$_1$ and LD$_5$ isoenzymes in their sensitivity to inhibition by excess of substrate *in vitro*, on the basis of which their different physiological functions were postulated, are not so marked at concentrations of substrate and enzyme corresponding to those believed to prevail *in vivo*.[60] Therefore, an alternative proposal has been made that abortive ternary complexes formed between lactate dehydrogenase, pyruvate, and the oxidised coenzyme, NAD$^+$, are responsible for enzyme inhibition within the cell, and that these complexes are more stable in the case of the LD$_1$ isoenzyme than with LD$_5$.[61] Although complete agreement has not yet been reached as to how far the different patterns of aerobic and anaerobic metabolism of tissues reflect the catalytic characteristics of their predominant isoenzymes of lactate dehydrogenase, these studies represent important attempts to give functional significance to variations in isoenzyme composition between tissues.

The MM and BB isoenzymes of creatine kinase differ to some extent in their quantitative catalytic properties, such as Michaelis constants for the various substrates of the forward or reverse reactions. As with lactate dehydrogenase, these catalytic differences between creatine kinase isoenzymes may provide indications of their physiological functions. For example, the MM isoenzyme may be better adapted to the rapid generation of ATP from creatine phosphate to provide energy for rapidly contracting skeletal muscle. The association of MM isoenzyme with white skeletal muscle fibres, and perhaps more specifically with their myofibrils, lends some support to this hypothesis.

The need for certain enzymic activities to be represented in both mitochondria and cytoplasm derives from the impermeability of the mitochondrial membrane to some metabolites, so that duplication of the reactions which generate these metabolites is necessary in order to make them available for mitochondrial or cytoplasmic metabolism. The existence of distinct mitochondrial and extra-mitochondrial isoenzymes potentially can provide in each of these subcellular compartments an enzyme form with catalytic characteristics appropriate to the substrate concentrations and direction of reaction prevailing in that location. An example is provided by the part played by the isoenzymes of malate dehydrogenase in the metabolism of acetyl-CoA. This compound is generated within the mitochondrion, the inner membrane of which is impermeable to it, but is used in the synthesis of fatty acids by enzymes located in the cytoplasm. A net transfer of acetyl-CoA from mitochondrion to cytoplasm is achieved by the intra-mitochondrial conversion of acetyl-CoA and oxaloacetate to citrate by the enzyme citrate synthetase. Citrate diffuses into the cytoplasm where acetyl-CoA is regenerated by the action of ATP-citrate lyase. The oxaloacetate formed in this reaction is reduced to malate by the cytoplasmic isoenzyme of malate dehydrogenase. Malate is further converted to pyruvate which can re-enter the mitochondrion, where oxaloacetate is again formed by stages including the oxidising action of mitochondrial malate dehydrogenase. This scheme suggests that consumption of malate is favoured in the mitochondrion and production of malate in the cytoplasm. The respective properties of the mitochondrial and cytoplasmic isoenzymes of malate dehydrogenase are consistent with this: the mitochondrial isoenzyme is inhibited by excess of oxaloacetate but not by malate, *i.e.*, it is better adapted for production of oxaloacetate from a relatively high concentration of malate, whereas the pattern of inhibition is reversed for the cytoplasmic isoenzyme, which is thus better adapated to catalyse the conversion of oxaloacetate to malate.

The metabolic significance of several families of isoenzymes remains obscure; in some cases, such as that of the non-specific phosphatases, this is because the place in metabolism of the particular catalytic activity represented by the isoenzymes is itself still uncertain. Differences between isoenzymes may result in differences in their biological half-lives and may contribute to the selective control of intracellular enzyme levels. It is conceivable that some isoenzymes are structural, rather than functional, variants, adapted for incorporation into the fabric of specific regions of the cell such as membranes. Some multiple forms of enzymes separable from tissue extracts may originate by post-genetic modifications, occurring either intracellularly or even artefactually during the extraction process. The latter is more likely to occur with those enzymes which are firmly bound to cellular structures.

The metabolic effects of enzyme variation which arise from the existence of allelic genes range from the undetectable to the catastrophic, when a key enzyme is so modified as to be unable to sustain its normal role in metabolism so that crippling disease or death is the result. Between these extremes, the possession of unusual isoenzymes may impose a handicap in the form of intermittent or chronic disease, or may only prove disadvantageous when the possessor is challenged by unusual circumstances, such as administration of a particular drug. The variety of effects which may be caused by genetic variation in a single enzyme is well exemplified by the many variants of glucose-6-phosphate dehydrogenase, each attributable to a different allele at a single gene locus. Possession of some variants apparently has no deleterious effects. Others carry with them susceptibility to haemolytic attacks when the subject is exposed to certain drugs or foodstuffs, such as the antimalarial drug Primaquine or the fava bean. The occurrence of still other variants is associated with mild or severe haemolytic anaemia.

The isoenzymes of glucose-6-phosphate dehydrogenase differ in one or more properties; net molecular charge, as demonstrated by differences in electrophoretic mobility, substrate affinity, pH dependence, and stability, but there is no clear correlation between the nature of the alteration in properties of the enzyme and the occurrence or nature of clinical symptoms. The relationship between allelic mutation and disease is shown decisively, however, when the product of the mutant gene has little or no catalytic activity. The list of diseases which arise from this extreme form of allelic mutation and in which the affected enzyme has been identified is now a very long one. It includes well-recognised disorders of amino-acid metabolism, and carbohydrate and lipid storage diseases. Certain of these diseases have an organ-specific distribution. This presumably occurs when the enzyme in question is the product of multiple gene loci which are expressed to different degrees in various tissues, so that mutation at only one locus produces a selective effect. For example, the phosphorylase which cleaves the α-1,4 glycosidic links of glycogen in muscle is a distinct isoenzyme from the analogous phosphorylase of liver. A mutation which decreases the activity of the former isoenzyme results in the accumulation of glycogen in skeletal muscle (McArdle's disease) while glycogen metabolism in liver is unaffected. These metabolic patterns are reversed in Hers' disease, in which the liver isoenzyme is deficient. In hereditary fructose intolerance, the failure to metabolise fructose is attributable to a specific deficiency of aldolase B, the predominant isoenzyme in normal liver.

Multiple Forms of Enzymes in Clinical Diagnosis

The clinical enzymologist attempts to infer the nature and extent of pathological changes in cells or tissues from a study of the activity and characteristics of enzymes in some readily obtainable sample, usually of blood serum. Although much useful information can be gained from measurements of enzyme activity alone, most of the enzymes which have been shown to be of diagnostic significance occur widely throughout the body, so that such measurements do not by themselves provide information as to the nature of the affected tissue in the absence of other evidence. The relative activities of several enzymes in a serum sample may reflect their relative activities in the tissue from which the enzymes originate; however, differential rates of release of enzymes from cells, or differential rates of removal from the circulation, may modify the pattern of enzyme activities in serum, while disease may be accompanied by changes in the rates of enzyme synthesis in the affected tissue.

The existence of multiple enzyme forms with tissue-specific patterns of distribution offers an alternative means of identifying affected organs by enzyme tests. For this approach to be potentially effective it is necessary that, as well as having a non-uniform distribution, the several enzymes should retain their identifying characteristics after the enzymes are released from their cells of origin. For isoenzyme studies to be useful in clinical practice, these characteristics should be sufficiently distinctive to allow them to be demonstrated reliably by relatively simple techniques. As well as the ability to trace a raised enzyme activity in serum to its origin in a particular organ or tissue, a second advantage of isoenzyme studies in diagnostic enzymology is their potential ability to detect pathological changes in the relative proportions of different isoenzymes in serum, even though the total enzyme activity may show minimal elevation or be within normal limits.

The extent to which these two advantages can be realised in diagnosis is closely related to the analytical methods employed and, as discussed in later chapters of this monograph, many improvements in methodology have resulted from the needs of diagnostic enzymology. Indeed, much of the impetus for the study of multiple forms of enzymes in general has originated in a desire to extend the value of enzyme tests in diagnosis.

Both the diagnostic advantages of increased tissue-specificity and sensitivity have been achieved by studies of the distribution of the components of several human isoenzyme systems in serum in disease, notably those of lactate dehydrogenase, creatine kinase, and alkaline phosphatase. Qualitative techniques have been used for many years to demonstrate the shift of the pattern of lactate dehydrogenase isoenzymes in serum towards that of heart or liver in patients with diseases of these organs (Fig. 1.8). Quantitative measurement of the MB isoenzyme of creatine kinase in serum is the most sensitive and specific enzymic

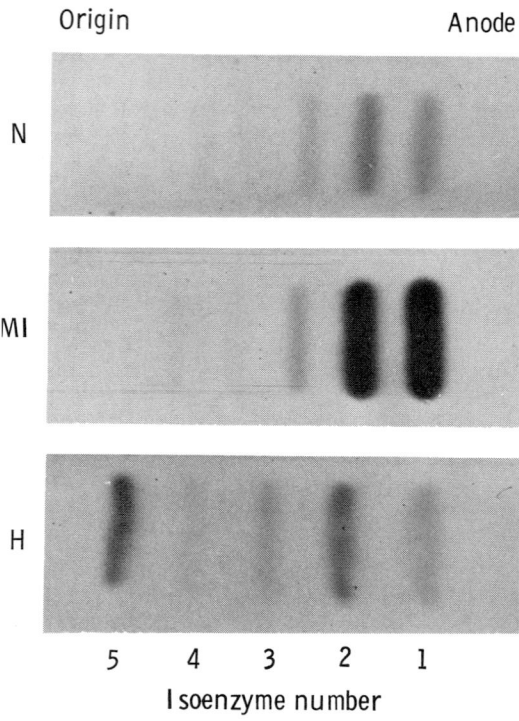

Figure 1.8 *Zones of lactate dehydrogenase activity in serum separated by electro-phoresis in agarose gel at pH 8.6. Activity has been demonstrated by coupling the enzymic oxidation of lactate and reduction of nicotinamide-adenine dinucleotide to the reduction of nitro-blue tetrazolium through the reduction and oxidation of phenazine methosulphate. Purple zones of reduced tetrazolium salt are produced at the positions of the isoenzymes, with intensities indicating their relative activities. Increased activities of isoenzymes 1 and 2 are seen in the serum from a patient with recent myocardial infarction*(MI) *and of isoenzyme 5 in serum from a case of acute infective hepatitis*(H), *compared with the normal distribution* (N)

indicator of damage to the myocardium yet devised. Specific measurement of the activity of the bone isoenzyme of alkaline phosphatase in serum is more sensitive than determination of total alkaline phosphatase activity in the investigation of osteoblastic bone disease (Fig. 1.9). The characteristic response of prostatic acid phosphatase towards certain inhibitors is used to detect metastases of prostatic carcinoma. Other diagnostic applications based on the distinctive properties of specific isoenzymes are also in use on a rather more limited scale.

When allelic variation of enzymes results in disease, as is the case for the inborn errors of metabolism, usually either no active enzyme is present, or the mutant isoenzyme is of such a low catalytic activity that demonstration of a low or absent total enzyme activity in an appropriate sample is pathognomonic, without the need for isoenzyme characterisation. Exceptions to this generalisation may arise in those conditions in which the existence of the mutant enzyme and its attendant consequences only become apparent on exposure to some environmental or therapeutic hazard. In these circumstances, isoenzyme characterisation may allow the probable response to particular conditions to be

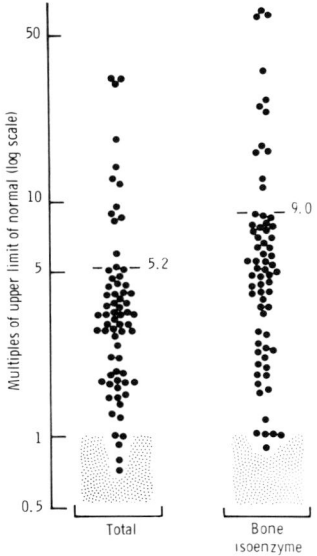

Figure 1.9 *Elevation of total and bone-type alkaline phosphatase activities in sera from patients with Paget's disease of bone. The nine-fold average elevation of the bone phosphatase levels is almost twice the average five-fold increase in total activity. Bone alkaline phosphatase activity was determined by a quantitative selective heat-inactivation method (see Chapter 3)*

anticipated and possible risks to be reduced. Also, since clinically similar diseases may result from different enzyme mutations, isoenzyme characterisation may help to distinguish between alternative diagnoses. Tay–Sachs disease and Sandhoff's disease (GM_2 gangliosidoses Type 1 and Type 2) are rather similar lipid-storage diseases, in which accumulation of the sphingolipid GM_2 in brain (Type 1) and also in viscera (Type 2) results in degeneration of the nervous system, blindness, and death. The deficient enzyme in each case is N-acetyl-β-D-hexosaminidase. Two forms of the enzyme, hexosaminidases A and B, are normally present in all tissues other than red blood cells. Hexosaminidase A has a greater net negative charge than B and therefore migrates further towards the anode on electrophoresis. It is also more readily denatured by heat than hexosaminidase B. An additional, more negatively charged, minor component, hexosaminidase S, is also detectable in normal tissues. The activity of hexosaminidase A is grossly reduced in Tay–Sachs disease whereas that of hexosaminidase B is normal or even increased in brain tissue. In Sandhoff's disease, on the other hand, both hexosaminidase A and B are deficient, although hexosaminidase S is more prominent.

A tetrameric structure has been postulated for hexosaminidase, in which the three isoenzymes represent homo- and hetero-polymers of two dissimilar subunits, α and β, such that hexosaminidases B and S are the homotetramers β_4 and α_4, respectively, and hexosaminidase A is the heterotetramer, $\alpha_2\beta_2$.[62] The isoenzyme changes in the two lipidoses are accounted for if the mutation responsible for Tay–Sachs disease affects the α-subunit, so that hexosaminidase A cannot be formed; in this case, the excess of β-subunits are available to form hexosaminidase B, which may be present in increased amounts. Similarly, if in Sandhoff's disease the formation of β-subunits is impaired, both hexosaminidases A and B are deficient but an excess of the α-subunit can give rise to an increased level of hexosaminidase S.

Many hereditary diseases are inherited in a recessive manner and their effects are only expressed fully in individuals who are homozygous for the disease-determining mutant allele (or in whom two different mutant alleles of similar effects have come together at the same locus). The amount of isoenzyme which is produced by the normal allele is usually sufficient to protect heterozygous individuals from the worst effects of enzyme deficiency. However, the normal isoenzyme may be quantitatively insufficient to prevent the appearance of some manifestations of metabolic abnormality in conditions of stress, *e.g.*, when an abnormal load of the compound whose metabolism is affected by the mutation is administered. Where the mutant allele produces little or no active enzyme, the average enzyme activity in heterozygotes is lower than that in homozygotes for the normal allele, although the degree of overlap between the two populations in this respect may be too great to allow a particular individual to be assigned with certainty to one or other group solely on the basis of measurments of total enzyme activity. If the mutant allele itself produces an active but altered isoenzyme, detection of the abnormal isoenzyme provides evidence of the heterozygous state. Although the practical applications of this approach are relatively limited so far, the importance of detection of carriers and appropriate genetic counselling is so great where fatal or crippling disease is concerned that attempts will probably be made to extend this aspect of isoenzyme studies.

Differences in Properties Between Multiple Forms of Enzymes

The structural differences between the multiple forms of an enzyme, whether arising at the level of the structural gene or as a result of post-genetic modification, may be accompanied by differences in properties which are clearly marked, or which are perceptible only by the most careful comparisons. Physico-chemical properties such as electrophoretic mobility, stability, or solubility, or catalytic characteristics such as reactivity towards substrate analogues or response to inhibitors, or properties belonging to both these categories may reflect differences between enzyme forms. Methods of isoenzyme characterisation are therefore designed to investigate a wide range of catalytic and structural properties of enzyme molecules. It is usually possible to make only limited deductions as to the nature of the underlying structural differences between isoenzymes which are responsible for the dissimilar properties observed with most of these methods. Equally, it is rarely possible reliably to predict the changes in catalytic and other properties which may result from specific structural alterations in enzyme molecules from current theoretical knowledge of the relationships between structure and function of proteins. The existence in living matter of a wide variety of multiple forms of enzymes provides an incentive for the development of methods for their investigation which will extend understanding of these relationships.

References for Chapter 1

1 Markert, C.L., and Møller, F., *Proc. Natl. Acad. Sci. U.S.A.*, 1959, **45**, 753.
2 IUPAC–IUB Commission on Biochemical Nomenclature, *J. Biol. Chem.*, 1977, **252**, 5939.
3 Hopkinson, D.A., Edwards, Y.H., and Harris, H., *Ann. Hum. Genet.*, 1976, **39**, 383.
4 Harris, H., and Hopkinson, D.A., *Ann. Hum. Genet.*, 1972, **36**, 9.
5 Donald, L.J., and Robson, E.B., *Ann. Hum. Genet.*, 1974, **37**, 303.

6   Lai, C.Y., Chen, C., and Horecker, B.L., *Biochem. Biophys. Res. Commun.*, 1970, **40**, 461.
7   Gazith, J., Schultze, I.T., Gooding, R.H., Womack, F.C., and Colowick, S.P., *Ann. N.Y. Acad. Sci.*, 1968, **151**, 307.
8   Moss, D.W., *Enzymologia*, 1970, **39**, 319.
9   Butterworth, P.J., and Moss, D.W., *Nature*, 1966, **209**, 805.
10 Samuelson, R.C., and Moss, D.W., *Clin. Chim. Acta*, 1978, **83**, 167.
11 Nayudu, P.R.V., and Hercus, F.B., *Biochem. J.*, 1974, **141**, 93.
12 Jacobson, K.B., *Science*, 1968, **159**, 324.
13 McKinley-McKee, J.S., and Moss, D.W., *Biochem. J.*, 1965, **96**, 583.
14 LaMotta, R.V., Woronick, C.L., and Reinfrank, R.F., *Arch. Biochem. Biophys.*, 1970, **136**, 448.
15 Wright, D.L., and Plummer, D.T., *Biochem. J.*, 1973, **133**, 521.
16 Ghosh, N.K., and Fishman, W.H., *Biochem. J.*, 1968, **108**, 779.
17 Moss, D.W., *Nature*, 1962, **193**, 981.
18 Biewenga, J., *Clin. Chim. Acta*, 1977, **76**, 149.
19 Ueda, M., Berk, J.E., Fridhandler, L., and Davis, J., *Clin. Chim. Acta*, 1971, **35**, 299.
20 Fridhandler, L., Berk, J.E., and Montgomery, K., *Clin. Chem.*, 1974, **20**, 26.
21 Qirbi, A.A., and Moss, D.W., *Clin. Chim. Acta*, 1975, **60**, 1.
22 Nagamine, M., and Ohkuma, S., *Clin. Chim. Acta*, 1975, **65**, 39.
23 Thorne, C.J.R., Grossman, L.I., and Kaplan, N.O., *Biochim. Biophys. Acta*, 1963, **73**, 193.
24 Davidson, R.G., and Cortner, J.A., *Science*, 1967, **157**, 1569.
25 Kitto, G.B., Wasserman, P.G., and Kaplan, N.O., *Proc. Natl. Acad. Sci. U.S.A.*, 1966, **56**, 578.
26 Schechter, A.N., and Epstein, C.J., *Science*, 1968, **159**, 997.
27 Kitto, G.B., Stolzenbach, F.E., and Kaplan, N.O., *Biochem. Biophys. Res. Commun.*, 1970, **38**, 31.
28 Richterich, R., Schafroth, P., and Aebi, H., *Clin. Chim. Acta*, 1963, **8**, 178.
29 Berg, T., and Blix, A.S., *Nature New Biol.*, 1973, **245**, 239.
30 Van Wijhe, M., Blanchaer, M.C., and St. George-Stubbs, S., *J. Histochem. Cytochem.*, 1964, **12**, 608.
31 Rosalki, S.B., in 'Proceedings of the Fourth Symposium on Current Research in Muscular Dystrophy', Pitman Medical, London, 1968, p. 348.
32 Fritz, P.J., Vesell, E.S., White, E.L., and Pruitt, K.M., *Proc. Natl. Acad. Sci. U.S.A.*, 1969, **62**, 558.
33 Wevers, R.A., Wolters, R.J., and Soons, J.B.J., *Clin. Chim. Acta*, 1977, **78**, 271.
34 Moss, D.W., *Enzyme*, 1975, **20**, 20.
35 Moss, D.W., and King, E.J., *Biochem. J.*, 1962, **84**, 192.
36 Boyer, S.H., *Ann. N.Y. Acad. Sci.*, 1963, **103**, 938.
37 Moss, D.W., Eaton, R.H., Smith, J.K., and Whitby, L.G., *Biochem. J.*, 1966, **98**, 32C.
38 Wootton, A.M., Neale, G., and Moss, D.W., *Clin. Chim. Acta*, 1975, **61**, 183.
39 Dreyfus, J.C., Demos, J., Schapira, F., and Schapira, G., *C.R. Acad. Sci.*, 1962, **254**, 4384.

40 Goto, I., Nagamine, M., and Katsuki, S., *Arch. Neurol.*, 1969, **20,** 422.
41 Hers, H.G., and Joassin, G., *Enzymol. Biol. Clin.*, 1961, **1,** 4.
42 Smith, M., Hopkinson, D.A., and Harris, H., *Ann. Hum. Genet.*, 1971, **34,** 251.
43 Whitaker, K.B., Whitby, L.G., and Moss, D.W., *Clin. Chim. Acta*, 1977, **80,** 209.
44 Walter, H., Welby, F.W., and Francisco, J.R., *Nature*, 1965, **208,** 76.
45 Holliday, R., and Tarrant, G.M., *Nature*, 1972, **238,** 26.
46 Wieme, R.J., and Herpol, J.W., *Nature*, 1962, **194,** 287.
47 Tzvetanova, E., *Enzyme*, 1971, **12,** 279.
48 Goldman, R.D., Kaplan, N.O., and Hall, T.C., *Cancer Res.*, 1964, **24,** 389.
49 Langvad, E., *Int. J. Cancer*, 1968, **3,** 17.
50 Schapira, F., Hatzfeld, A., and Weber, A., in 'Isozymes, III Developmental Biology', ed. Markert, C.L., Academic Press, New York, 1975, p. 987.
51 Criss, W.E., *Cancer Res.* , 1971, **31,** 1523.
52 Fishman, W.H., Inglis, N.R., Greene, S., Anstiss, C.L., Gosh, N.K., Reif, A.E., Rustigian, R., Krant, M.J., and Stolbach, L.L., *Nature*, 1968, **219,** 697.
53 Greene, P.J., and Sussman, H.W., *Proc. Natl. Acad. Sci. U.S.A.*, 1973, **70,** 2936.
54 Moss, D.W., Campbell, D.M., Anagnostou-Kakaras, E., and King, E.J., *Pure Appl. Chem.*, 1961, **3,** 397.
55 Hodson, A.W., Latner, A.L., and Raine, L., *Clin. Chim. Acta*, 1962, **7,** 255.
56 Cox, R.P., Elson, N.A., Tu, S., and Griffin, M.J., *J. Mol. Biol.*, 1971, **58,** 197.
57 Horecker, B.L., in 'Isozymes, I Molecular Structure', ed. Markert, C.L., Academic Press, New York, 1975, p. 11.
58 Cahn, R.D., Kaplan, N.O., Levine, L., and Zwilling, E., *Science*, 1962, **136,** 962.
59 Wilson, A.C., Cahn, R.D., and Kaplan, N.O., *Nature*, 1963, **197,** 331.
60 Vesell, E.S., and Pool, P.E., *Proc. Natl. Acad. Sci. U.S.A.*, 1966, **55,** 756.
61 Kaplan, N.O., Everse, J., and Admiraal, J., *Ann. N.Y. Acad. Sci.*, 1968, **151,** 400.
62 Beutler, E., and Kuhl, W., *Nature*, 1975, **258,** 262.

2

Separation of Multiple Forms of Enzymes

The existence of multiple forms of enzymes and the nature and extent of the differences between them can be fully confirmed and determined only when methods are available for their separation and individual purification. Because of the general similarities in properties between the multiple forms, these methods must be of considerable resolving power. It is not surprising, therefore, that the development of knowledge of the existence and nature of enzyme heterogeneity has accompanied, and has depended on, developments in methods of resolving mixtures of similar proteins. Advances in ion-exchange chromatography and improvements in the resolving power of zone electrophoresis have been particularly significant in the study of isoenzymes.

Zone Electrophoresis

Analysis of mixtures by zone electrophoresis is the most useful single technique in the study of multiple forms of enzymes. The potential ability of zone electrophoresis to detect small differences in composition between protein molecules is very great. About one-third of all substitutions of a single amino-acid residue for another in a polypeptide chain are likely to involve a change in the ionic charge of the residue at that position, *e.g.*, by substituting a basic amino acid for one which is neutral or acidic, and will therefore result directly in an alteration in the net molecular charge of the polypeptide. However, substitution of one residue for another of different ionic charge is not the only way in which differences in the overall charge of a protein molecule may be produced. Since the higher levels of protein structure are determined by the primary structure (*i.e.*, the amino-acid sequence), substitution of one amino acid for another may cause a change in molecular conformation such that ionisable residues different from those in the unmodified molecule are exposed to the surrounding medium. While single amino-acid substitutions probably account for most allelic variants of enzymes, other modifications of structure can occur, such as deletion or repetition of a section of the primary structure, with a probable alteration in molecular charge. Thus, although some variations in protein structure arising from the existence of multiple gene loci or alleles may not involve a change in molecular charge and so will escape detection by electrophoresis, these cases probably make up a smaller proportion of all the possible changes in primary structure than a consideration of the ionisation characteristics of amino acids alone would suggest. Similarly, many of the post-genetic modifications which may give rise to multiple forms of enzymes, such as side-chain cleavage, addition of non-protein components, or aggregation, will result in altered molecular charge.

As an analytical procedure, zone electrophoresis requires only small samples, when combined with suitably sensitive methods of detecting the separated isoenzyme zones. A further advantage of the technique is that the separation achieved is generally unaffected by the degree of purity of the sample being separated. Electrophoresis is therefore particularly valuable for investigating the isoenzyme composition of unpurified extracts of cells or tissues, as well as serum or other body fluids, and in consequence it has been used extensively for genetic and clinical studies. Exceptions to this generalisation are occasionally encountered, however, such as differences in the proportions of minor zones of alkaline phosphatase activity seen with alternative methods of extraction of the enzyme from liver,[1] or the changes in mobility of isoenzymes of creatine kinase which result from incubation in serum but not in saline.[2] Purification may sometimes remove attached ions or small molecules, or dissociate isoenzyme complexes, so altering electrophoretic mobility. Compared with some of the non-separative methods of isoenzyme characterisation discussed later, electrophoresis also offers a high degree of certainty in establishing the identities of particular isoenzymes present in the original mixture.

Supporting Media in Zone Electrophoresis.—Although some of the earliest investigations of the electrophoretic heterogeneity of enzymes were carried out by paper electrophoresis,[3,4,5] the functions of paper as a supporting medium have now largely been taken over by various forms of cellulose acetate.

(*a*) *Cellulose acetate.* Compared with filter-paper as a supporting medium, cellulose acetate membranes offer much reduced adsorption of protein zones, so that clearer separation and lower background staining are obtained.[6] Electrophoretic separation is rapid, resolution of isoenzyme zones of diagnostic interest in serum typically being obtained in 1 h or less with potential differences of the order of 25 V cm^{-1} applied across the ends of a 10 cm strip, with currents of about 0.5 mA cm^{-1} width.

Thin cellulose acetate membranes are not very absorbent, so that the volume of isoenzyme solution (*e.g.*, serum) which can be applied without excessive spreading of the applied zone is limited to about 3 μl cm^{-1} of the width of the strip. The volumes of reagent solutions required to demonstrate isoenzyme zones after electrophoresis on cellulose acetate membranes are small, but, because of the low liquid-absorbance of the material, the staining reagents may have to be applied by placing filter-paper or cellulose acetate moistened with them in contact with the electrophoretic strip, or by incorporating the reagents in an overlay of agar gel, to prevent elution or spreading of isoenzyme zones. Recovery of isoenzyme zones by elution from cellulose acetate membranes is good, although the total amount of an isoenzyme which can be obtained in this way is limited by the small sample size. Cellulose acetate can readily be made transparent for densitometric scanning.

Some of the disadvantages of thin cellulose acetate membranes, such as the difficulty of wetting the strip evenly with buffer solution or of applying a uniform line of sample solution, and of handling the brittle, dry strips, are reduced in other forms of the material. These include more porous forms and cellulose acetate backed with clear, flexible plastic.

Electrophoresis on cellulose acetate is widely used in diagnostic applications of isoenzyme studies, in which its speed and simplicity are particularly advantageous[7,8,9] (Fig. 2.1).

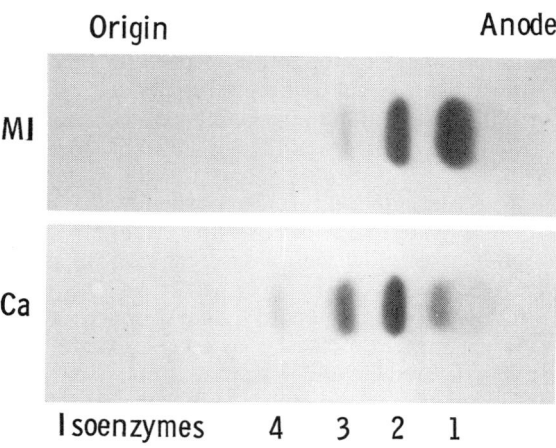

Figure 2.1 *Separation of lactate dehydrogenase isoenzymes in serum by electrophoresis on cellulose acetate at pH 8.6. The method of staining was as for Fig. 1.8. The characteristic increase in the activities of isoenzymes 1 and 2 is seen in the myocardial infarction specimen (MI). The serum (Ca) from a patient with lung cancer shows a relative increase in isoenzyme 3 compared with normal sera, with traces of isoenzymes 4 and 5. Total activity is about twice the upper normal level*

(b) *Agar and agarose gels*. These media share some of the advantages of cellulose acetate in that they have low non-specific adsorption of proteins and allow rapid separation of isoenzyme zones. In addition, dried films of the gels are transparent, so that direct photometric or densitometric scanning is possible, while passive diffusion is more restricted in the gels than on cellulose acetate, which contributes to sharper, more clearly resolved isoenzyme zones. A sample-volume of about 10 μl cm^{-1} width can be applied to an agar gel 1—2 mm thick, either introduced into a slit in the gel or soaked into a strip of filter-paper placed on its surface. The composition of the gel is usually 0.5—1.0% (m/V) in an appropriate buffer. Separation of isoenzyme zones in serum takes 2—3 h or less, with an applied voltage of the order of 35 V cm^{-1} and currents of about 0.5 mA cm^{-1} width. Development of isoenzyme zones after electrophoresis is usually carried out by overlaying the electrophoresis gel with a second layer of gel containing the staining reagents.

Variability of results of agar-gel electrophoresis can result from the use of batches of agar of different purities.[10] The use of purified agarose, the straight-chain polymer fraction of agar, removes much of this variability of properties, and also much of the considerable electro-endosmosis which is typical of agar-gel electrophoresis because of the acidic sulphate and carboxylate groups present in the agaropectin fraction.

Electrophoresis on agar or agarose gels has been used in the study of many isoenzyme systems, including among others lactate dehydrogenase[11-14] (Fig. 1.8), alkaline phosphatase[15,16] and other esterases, aspartate aminotransferase,[17] and creatine kinase.[18,19,20] A detailed account of the electrophoresis technique has been given by Wieme.[21] Agar- or agarose-gel electrophoresis is also particularly useful in conjunction with various immunochemical procedures.

(c) *Starch and polyacrylamide gels*. Supporting media such as cellulose acetate membranes or agar gel impose no selective restrictions on the passage through them of molecules of different sizes. In gels formed from partially-hydrolysed starch or by polymerisation and cross-linking of acrylamide, however, the pores approximate to macromolecular dimensions, so that larger protein molecules or complexes are retarded during electrophoresis compared with smaller molecules of similar charge. Restricted diffusion of protein molecules in these gels also helps in maintaining narrow zones during electrophoresis and the subsequent staining process. Starch-gel electrophoresis has occupied a particularly important place in the study of multiple forms of enzymes and continues to do so, as is demonstrated by the extensive compilations of methods which have been published,[22,23] although it has been replaced to some extent by electrophoresis in polyacrylamide gels in recent years.

Electrophoresis in horizontal starch gels cast in plastic trays was first described in 1955 by Smithies.[24] The gel thickness is from 4 to 7 mm; thicker gels are difficult to keep cool during electrophoresis. Samples are applied by absorbing them into small pieces of thick filter-paper which are then inserted into vertical slots cut in the gel, or are introduced into slots in the gel as a slurry with starch grains. These alternatives prevent electro-decantation which takes place when sample solution alone is introduced into a slot in the gel, distorting the vertical boundaries of the zones. When rectangles of Whatman 3MM filter-paper are used to carry the samples, volumes of up to 20 μl cm^{-1} width of gel can be applied. In a later modification of the technique, electrophoresis is carried out with the gel mould held in a vertical position, with the anode at the bottom.[25] This allows liquid samples to be inserted into horizontal slots in the gel, without the need for any supporting medium, since there is no possibility of electro-decantation. The slots are cast into the gel around a comb-shaped former with square 'teeth' about 1-cm wide separated by gaps of 3 mm. The use of free liquid samples eliminates possible blurring of the zones by adsorption to paper or starch grains, so that the resolution of

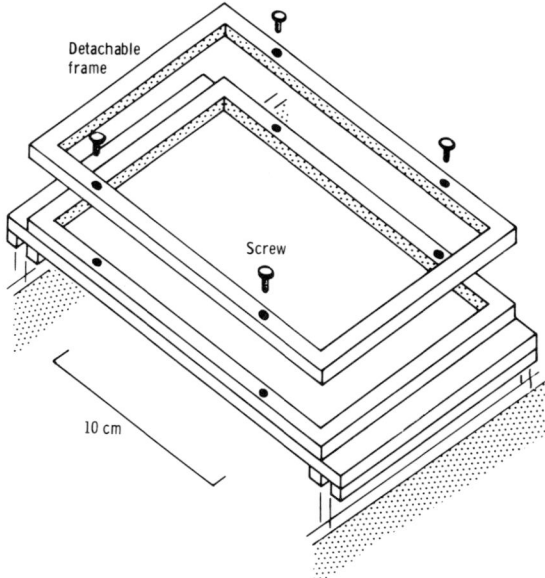

Figure 2.2　*Isometric drawing of the construction of a tray used for horizontal starch-gel electrophoresis. The material is Perspex sheet 6 mm thick. The grooves on the underside of the base locate the assembled tray on the edges of the electrode vessels, parts of which are shown*

closely spaced zones may be improved. However, the simpler, horizontal arrangement has been found to give excellent resolution of enzyme zones in the author's laboratory, in which Perspex trays of the construction shown in Fig. 2.2 have been used for many years.

To be suitable for making gels, potato starch must be hydrolysed partially by heating with concentrated HCl (1 vol. to 100 vol. acetone) for $\frac{1}{2}$ to $2\frac{1}{2}$ h. If hydrolysis is insufficient, the starch will not form a gel when heated in buffer, but gels formed from over-hydrolysed starch are too weak for use.[26] Suitable hydrolysed starches can be obtained commercially, *e.g.*, from Connaught Medical Research Laboratories, Toronto, Canada. The gel is prepared by heating a suspension of the hydrolysed starch (10—15% m/V) in buffer solution in a round-bottomed flask over a medium bunsen flame with constant swirling. The mixture at first becomes very viscous, then fluid and more transparent with a few bubbles appearing. Suction is applied from a filter-pump while the mixture is still swirled. The vigorous evolution of gas is quickly followed by slower formation of larger bubbles. Heating and suction are then removed, and a mobile liquid gel is obtained which is poured rapidly into the tray (Fig. 2.2) and allowed to cool and set. Before the gel is poured into the tray, a loose-fitting glass plate is placed in the bottom to facilitate later removal of the gel and filter-paper wicks are also inserted at the ends to connect the gel to the electrode tanks. The surface of the gel is sealed against evaporation by thin plastic film and a voltage gradient of about 6 V cm^{-1} applied for 16—18 h, or up to 20 V cm^{-1} for a shorter period. Since the protein zones are distorted near its upper and lower surfaces, the gel is sliced in two horizontally along its length after electrophoresis, the upper and lower halves are separated, and the inner cut surfaces are stained. Slicing is facilitated by removal of the upper rim of the tray, which exposes the top half of the gel, the lower rim serving as a guide.[27] A thin, stainless-steel wire held taut between the hands is drawn through the gel after first removing the filter-paper sample inserts.

Buffer solutions covering a wide range of pH values and compositions have been used in starch-gel electrophoresis. The molarity of the gel buffer is usually about one-tenth of that in the electrode compartments, *e.g.*, 0.03 mol l^{-1} borate compared with 0.3 mol l^{-1}.[24] 'Discontinuous' buffer systems, in which the electrode buffer and the gel buffer have different compositions, are particularly useful. In the first such system to be described[28] the gel is made in a buffer of Tris (0.076 mol l^{-1}) and citrate (0.005 mol l^{-1}), pH 8.6, while the electrode vessels contain borate (0.3 mol l^{-1})–NaOH, pH 8.5. Discontinuous buffer systems give improved separations because of the passage through the gel during electrophoresis of a front of increased voltage gradient as, for example, borate ions replace citrate ions, causing sharpening of the trailing edges of the protein zones. The junction of the borate and Tris–citrate buffers can be seen as a yellow line advancing through the gel towards the anode, which is a useful marker of the progress of electrophoresis.

Starch gels consist of a system of long, interwoven, branching molecules, in which the size of the pores is reduced as the concentration of starch in the gels is increased. The migration of proteins through the gel is inversely proportional to starch concentration in the range from 11.4 to 15.6% m/V, retardation in more concentrated gels being greater for larger than for smaller protein molecules.[29] This effect has been used to estimate molecular sizes of proteins, but this is more easily done by the techniques of gel filtration or electrophoresis in polyacrylamide gels of graded pore sizes described later.

The applications of starch-gel electrophoresis to studies of isoenzymes are numerous.[22, 23, 30, 31] They include separations of systems of isoenzymes which do not involve differences in molecular size where the sieving properties of the gel do not come into play but where the technique has been chosen for its good resolution and compact isoenzyme zones (*e.g.*, compare the separations of lactate dehydrogenase isoenzymes on different media, in Figs. 1.8, 2.1, and 3.2), as well as separations in which components differing in size are present (Fig. 2.3). Starch-gel electrophoresis can also be combined with electrophoresis on filter-paper or other supporting media to give two-dimensional separations.[26] This allows differences between multiple forms of enzymes which are due to differences in net charge to be distinguished

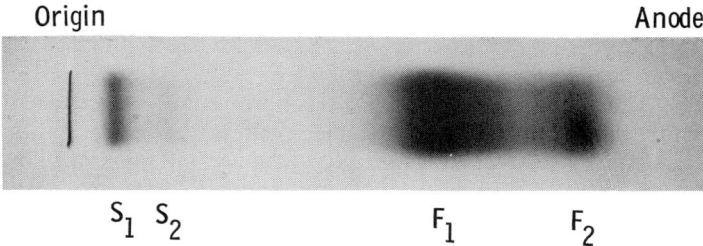

Origin   Anode

S_1 S_2 F_1 F_2

Figure 2.3 *Zones of alkaline phosphatase activity in an extract of human small intestine, separated by horizontal starch-gel electrophoresis in a discontinuous Tris–citrate–borate buffer system at pH 8.6. The active enzyme zones were rendered visible by incubating the gel after electrophoresis in a solution of 1-naphthyl phosphate (5 mmol l⁻¹) in carbonate–bicarbonate buffer, pH 10, then, after washing briefly, immersing it in a solution of Fast Blue B salt (tetra-azotised o-dianisidine) to produce zones of purple azo dye at the sites of 1-naphthol production. The gel was preserved in a mixture of methanol, acetic acid and water(5 : 1 : 5 V/V). The slow-moving zones, S_1, S_2, and similar zones in extracts of other tissues are retarded in starch gel because of their greater molecular size compared to the main zones, F_1, F_2*

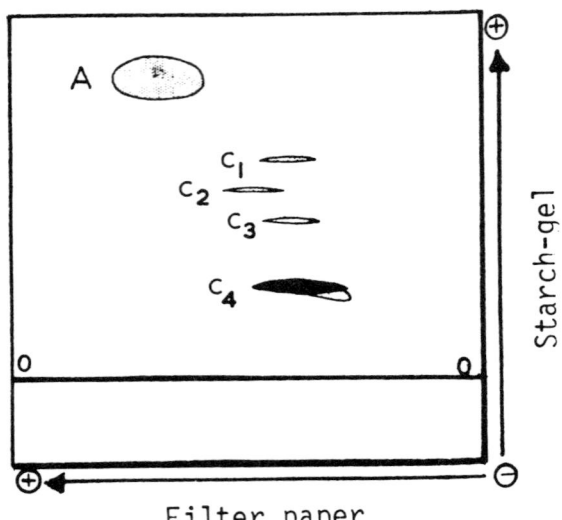

Figure 2.4 *Two-dimensional zone electrophoresis on filter-paper (first dimension) and in starch gel (second dimension) at pH 8.6 of components* C_1, C_2, C_3, *and* C_4 *of human serum cholinesterase. The zones are not fully resolved by paper electrophoresis but are well separated in starch gel because of their differences in molecular size. A, serum albumin; 0—0, line of insertion of paper-electrophoresis strip into gel.* [Reproduced, with permission, from: H. Harris, D. A. Hopkinson, and E. B. Robson, *Nature* (Macmillan), 1962, **196**, 1296]

from those involving differences in size, *e.g.*, among the multiple forms of cholinesterase in human serum[32] (Fig. 2.4).

Polyacrylamide gels for zone electrophoresis are formed by polymerisation of a mixture of acrylamide with a small proportion of bisacrylamide (N,N'-methylene-bisacrylamide). Polymerisation is brought about by the presence of free radicals, most conveniently generated by adding ammonium persulphate. TEMED (N,N,N',N'-tetramethyl-1,2-diaminoethane) is added as a catalyst. The process is inhibited by excess of oxygen. An alternative source of free radicals is photolysis by ultraviolet light of small amounts of riboflavin added to the monomer solution. Gel formation is due to the cross-linking by bisacrylamide of long-chain polyacrylamide molecules to form macromolecules of parallel chains which together make up the gel. Pore size depends on the total acrylamide concentration and also on the proportion of bisacrylamide present, becoming smaller as the total acrylamide concentration is increased, but reaching a minimum value with respect to bisacrylamide concentration when the latter constitutes about 5% of the total acrylamide concentration.[33] Useful concentrations for preparing gels for isoenzyme separations are of the order of 5—6% m/V total acrylamide, of which 5% m/m is bisacrylamide.

Gels prepared in this way have virtually no charged groups, so that the electro-endosmosis or adsorption of protein molecules due to ionic effects which may result from the small number of carboxylate groups present in starch gel do not occur. Symmetrical protein zones are therefore obtained by electrophoresis in polyacrylamide gels. However, aromatic residues in proteins may interact with polyacrylamide to some extent. Polyacrylamide gels are transparent and, being tough but flexible, are more easily handled than starch gels. However, the acrylamide monomers are toxic and appropriate precautions should be taken in their use.

Polyacrylamide gels cast in narrow glass tubes arranged vertically with the anode downwards during electrophoresis were introduced by Ornstein and Davis.[34, 35] Separation takes place in a small-pore gel, above which is a 'spacer gel' of larger pore size (*i.e.*, lower acrylamide concentration), made in a buffer solution of lower molarity than that in the separation gel. The sample is applied in a further superimposed gel layer. A buffer of different composition is used in the electrode vessels.[35] In this arrangement, the protein components migrate downwards to become concentrated at the interface between the larger- and smaller-pore gels, before entering and being separated in the latter. Thus, the protein zones enter the smaller-pore gel as narrow discs, which have given the technique its name of 'disc electrophoresis', and extremely sharp, well-resolved bands are obtained. Concentration is due to the Kohlrausch effect, by which a steep potential gradient is formed at the sharp boundary between ions of lower mobility ('trailing ions') and faster, 'leading' ions in an electrophoretic system. With the arrangement of spacer gel and separating gel, protein ions do not enter the separating gel until the front between different buffer ions has passed through, so that zone broadening before separation begins cannot occur.

Electrophoresis in separate tubes has disadvantages when the isoenzyme components of different samples are to be compared, since slight differences in migration can occur from tube to tube, however carefully the conditions are standardised for each tube. Systems in which samples can be separated side by side in thin slabs or sheets[36, 37, 38] are preferable for some applications, such as the comparison of isoenzyme zones in serum samples. Furthermore, the three-gel system of Ornstein and Davis is rather complicated to set up, and zones of nearly equal sharpness can be obtained in a single gel with a continuous buffer system, if the sample is applied as a concentrated solution of lower conductivity than the buffer itself.[39]

Analytical electrophoresis in thin slabs of uniform pore-size polyacrylamide gels, made in a rectangular glass cell modified from the apparatus

(a)

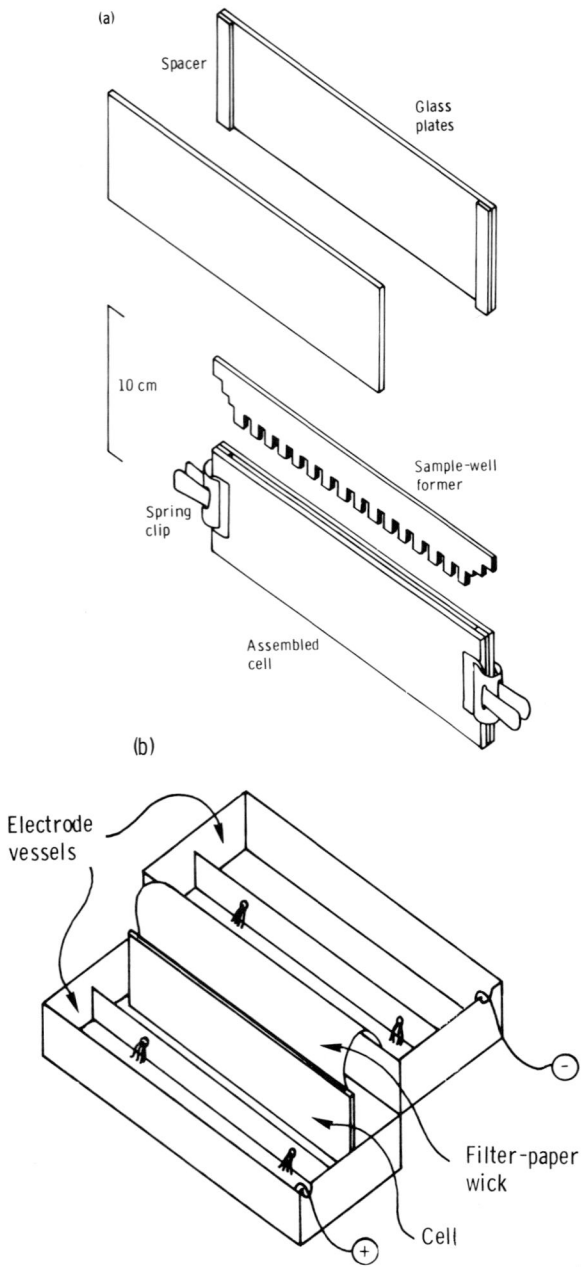

Spacer

Glass
plates

10 cm

Sample-well
former

Spring
clip

Assembled
cell

(b)

Electrode
vessels

Filter-paper
wick

Cell

−

+

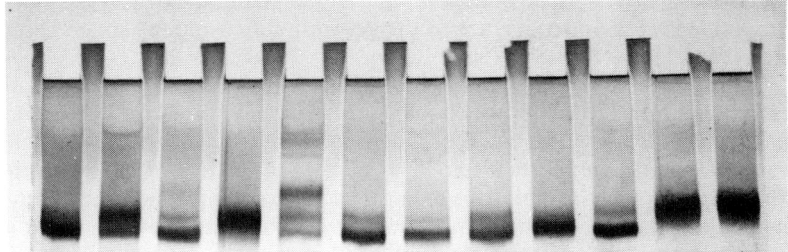

Figure 2.6   *Separation of zones of alkaline phosphatase activity in serum samples by polyacrylamide-gel electrophoresis at pH 9.5 in the apparatus shown in Fig. 2.5. Staining for enzyme activity was carried out by immersing the gel in a solution of sodium 1-naphthyl phosphate (50 mg) and Fast Blue BB salt (diazotised 4'-amino-2',5'-diethoxybenzanilide; 70 mg) in the Tris–borate– MgCl₂ buffer (70 ml) used for electrophoresis, then, after the bands were visible, washing the gel in methanol–acetic acid–water (5:1:5 V/V). The most-anodal zone of liver phosphatase is present in most samples (e.g., the third and tenth from the left), the more diffuse bone-phosphatase zone is prominent in samples 4, 11, and 12, while the slower zone of the intestinal isoenzyme is present in sample 5. Some non-specific staining can also be seen. The sample wells are at the top and the anode at the bottom*

described by Akroyd,[38] is used in the author's laboratory (Fig. 2.5). The cell is formed of two glass plates, to one of which a 2-mm thick glass spacer is glued at each end. Before use, the plates are cleaned with detergent solution and rinsed. The spacers are covered with a thin film of silicone grease to make a complete seal when the plates are clamped together with strong spring clips. The cell is held upright in a tray and the bottom is sealed with agar gel (1% m/V agar dissolved with boiling in a solution of NaCl, 0.04 mol l⁻¹) to a depth of about 5 mm. After the agar gel has set, a solution of acrylamide monomers, TEMED, and ammonium persulphate, in the proportions given above and in a suitable buffer, is poured into the cell, nearly filling it. The top is closed with the Perspex comb, around which the sample wells are moulded[25,37] and which also excludes air while the gel is setting. After the polyacrylamide gel has set the agar gel outside the cell is removed, leaving the agar-gel seal inside, the comb is removed gently, and the cell is set upright in the

Figure 2.5   (a) *Isometric drawing of the construction of a glass cell for vertical zone electrophoresis in polyacrylamide gel, based on the design of Akroyd.[38] The thickness of the gel is 2 mm. The sample-well former is made from Perspex of this thickness and fits closely within the assembled cell*
(b) *General arrangement drawing of the assembled apparatus during electrophoresis*

anode tank of the electrophoresis apparatus. Unreacted acrylamide is washed from the upper surface of the gel with buffer and the space above the gel is then filled with buffer solution.

Samples (10—15 μl per well, mixed with sucrose solution) are introduced into the wells from a capillary pipette with its tip held under the buffer surface. The sucrose solution increases the density of the samples, so that they form a thin layer on the gel surface, and reduces the conductivity of the samples, so aiding zone sharpening. The space above the gel is connected to the cathode tank by a filter-paper wick soaked in buffer solution. A potential gradient of about 4 V cm^{-1} is applied until the samples have entered the gel and is then increased to about 8 V cm^{-1} until separation is complete (3—4 h for isoenzymes of clinical interest in serum; Fig. 2.6).

A large variety of buffer systems covering a wide range of pH values can be used in polyacrylamide gel electrophoresis, including discontinuous systems combining buffers of different compositions to promote zone sharpening.

The reproducibility of properties of polyacrylamide gels and the readiness with which their pore sizes can be varied has led to the introduction of electrophoresis in gradient-pore gels for the estimation of molecular weights.[40] Gels in which the acrylamide concentration ranges from 4—24% m/V are prepared by filling a vertical gel mould with solutions of high and low concentrations of acrylamide from a gradient-forming device. Linear or concave gradients of acrylamide concentration as a function of distance along the gel can be produced in this way. Electrophoresis is carried out with the origin at the low-concentration border of the gel and with the direction of migration towards the regions of higher concentration. The protein zones pass through progressively smaller and smaller pores, until an equilibrium is reached when the molecules are unable to make any further progress through the gel and migration ceases. This usually requires more than 24 h with mixtures of proteins covering the range of sizes found in serum. Zone-sharpening occurs because of the tendency of the trailing edges of zones to overtake the leading edges as the latters' migration ceases.

In gradient-pore electrophoresis, therefore, proteins move into the gel under the influence of their net molecular charges, but their final positions in the gel and the separation between them are determined by their relative molecular sizes. The gel can be calibrated by the introduction of proteins of known molecular weight, with which the proteins being characterised can be compared. Pore size is inversely proportional to the square root of gel concentration, so that the change in pore size with distance is not a linear function of the distance along the gel in a gel prepared with a linear concentration gradient, successive equal increments of gel concentration producing decreasing increments of pore size. Therefore, the smaller protein molecules have to travel a relatively greater distance through a linear-gradient gel than larger molecules before reaching the limiting pore size at which migration ceases, the last part through a region in which their movement is very much impeded. Consequently, long electrophoresis periods are needed to reach equilibrium. This problem can be overcome by the use of gels with a concave gradient of concentration with respect to length, *i.e.*, in which pore size is linearly related to distance along the gel.

The migration of non-spherical molecules is affected by their orientation with respect to the pores of the gel, and the technique of gradient-pore electrophoresis is unsuitable for the separation of such molecules. However, enzymes are typically globular proteins to which this limitation does not apply. The zone-sharpening effect of electrophoresis in gradient-pore gels has been used to provide improved electrophoretic separations in runs of about the same durations as are used in electrophoresis in gels of uniform pore size. When the components being separated are of nearly equal sizes, the electrophoresis-time for maximum resolution is critical, since too short a run does not cause sufficient retardation of the leading edges to bring about significant compaction of the zones, while too protracted electrophoresis reduces the separation between them.

Differences in molecular size can also be made the dominating factor in separations in polyacrylamide gel by first treating the mixture of proteins to be analysed with sodium dodecyl sulphate (SDS), then adding a low concentration of SDS (as little as 0.1—0.2% m/V) to the buffer solutions used for electrophoresis.[41] SDS combines with polypeptides in the ratio of approximately 1.4 g SDS per g of protein, corresponding to about one molecule of SDS for each pair of amino-acid residues. Consequently, whatever their original charge properties, different polypeptides thus treated acquire essentially equal ratios of negative charge to mass, because of the attached SDS molecules. When submitted to electrophoresis in polyacrylamide gels, in which resistance to migration varies directly with molecular size, differences in rates of migration of the treated polypeptides reflect their relative sizes. Electrophoresis of various SDS-treated proteins in polyacrylamide gel containing SDS has shown a rectilinear relationship between electrophoretic mobility and the known molecular weights of monomeric proteins, or subunits of polymeric proteins, plotted on a logarithmic scale. Estimates of molecular weight obtained by comparing known and unknown polypeptides under these conditions are probably accurate to $\pm 10\%$ in the range from 15 000 to 100 000.[42]

Reaction with SDS denatures proteins, dissociating polymeric molecules and unfolding their constituent subunits. It may be necessary to provide additional denaturing conditions, such as heat, concentrated urea solutions, or sulphydryl reducing agents, to ensure that the reaction of proteins with SDS is complete, and to prevent aggregation of polypeptides during electrophoresis by adding suitable reagents to the buffer solutions (*e.g.*, β-mercaptoethanol, to prevent re-formation of disulphide bonds). Because denaturation of enzymes destroys their catalytic activity, enzyme zones separated by SDS electrophoresis cannot be visualised by methods which make use of the reaction catalysed by the enzyme, but only by non-specific methods such as the use of protein stains. Therefore, in contrast to gradient-pore electrophoresis, SDS electrophoresis cannot be used to fractionate impure mixtures of isoenzymes, even if they differ in molecular size. However, the SDS technique has proved valuable in studies on purified isoenzymes in determining and comparing their subunit compositions.

Dissociation of proteins into their component subunits can usually be achieved by treatment with concentrated solutions of urea or guanidine hydrochloride, followed by electrophoresis in buffers containing these reagents, to separate the subunits. Starch gels can be made in buffers containing urea up to 8 moles l^{-1} and, although setting time is much prolonged, the gels are stronger and more transparent than gels from which urea is absent. The properties of polyacrylamide gel are unaffected by the presence of high concentrations of urea.

Visualisation and Quantitation of Isoenzyme Zones.—One of the particular advantages of zone electrophoresis in the study of multiple forms of enzymes is the ability in many cases to make the separated enzyme zones visible *in situ* in the supporting medium, because of their catalytic activity. The methods used for this purpose are usually derived and adapted from those employed by histochemists to locate enzymes in tissue sections and, as in histochemistry, the presence of enzymes of almost every class can be demonstrated by a suitable choice of substrates or coupled reaction sequences, although ligases and isomerases present difficult methodological problems. The factors which influence the design of methods for locating enzyme zones after electrophoresis are also similar in some respects to those which operate in histochemistry; notably, the need for distinctively coloured reaction products, which are either insoluble or of low diffusibility so that they remain at the sites of enzyme action. The latter requirement, so stringent in histochemistry, can be relaxed to some extent in isoenzyme electrophoresis since a small loss of resolution may be offset by gains of sensitivity or convenience. Similarly, the precautions taken by the histochemist to ensure that reactions competing with, or analogous to, the one under study do not interfere are usually made less necessary by electrophoresis, especially when the enzyme zones are compared under several different conditions of separation.

Reagent solutions for demonstrating enzyme activity can be applied directly to the electrophoresis support. This is the technique usually adopted when the supporting medium is starch or polyacrylamide gel. Fixation of enzyme zones before staining is to be avoided, since the enzymes may be inactivated, but the restricted diffusion in gel media generally prevents significant loss of enzymes into the solution. However, this property of gels may reduce the access of staining reagents to the enzyme zones and therefore the sensitivity of the staining process; this is more likely to happen when polyacrylamide gel is used which, unlike starch gel, is typically not sliced before staining, and when the staining mixture contains enzymes as components of coupled reaction systems. The presence of reagent enzymes at the sites of separated enzyme zones can be ensured by adding them to the electrophoresis buffer solutions.[43] When electrophoresis is carried out on thin support media, such as cellulose acetate or agar gel, the ease with which enzyme zones may be eluted into solution requires the use of an overlay of thick filter-paper or a superimposed layer of agar gel containing the reaction mixture.[21]

From the wide range of methods which have been described for the location of specific isoenzyme zones after electrophoresis,[22, 23, 30] certain principles have emerged which are of general applicability. Coupled oxidation–reduction reactions, by which conversion of the coenzyme NAD^+ to NADH by the action of lactate dehydrogenase was rendered visible by reduction of a tetrazolium salt to an intensely-coloured, insoluble formazan, were introduced by Markert and Møller to locate the isoenzymes of this enzyme after starch-gel electrophoresis.[44] In the procedure originally described, pyruvate formed by the oxidation of lactate was trapped as a hydrazone by adding hydrazine to the mixture. Cyanide has been used as an alternative way of trapping pyruvate, but addition of neither reagent is necessary. Diaphorase and methylene blue, the intermediate electron carriers of the original system, have also been replaced by phenazine methosulphate (PMS; methylphenazonium methosulphate). Of the available tetrazolium salts, nitro blue tetrazolium [NBT; 2,2'-di-*p*-nitrophenyl-5,5'-diphenyl-3,3'-(3,3'-dimethoxy-4,4'-diphenylene)-ditetrazolium chloride] is probably the most useful. The diformazan to which it is reduced is intensely coloured and insoluble, not only in the buffer solutions in which the enzyme zones react, but also in the acidic methanol solutions used to dehydrate and preserve starch gels.

A general staining reagent for the demonstration of reduction of NAD^+ (or the similar coenzyme $NADP^+$) can therefore be made by adding NBT (30—50 mg 100 ml^{-1}) and PMS (1—2 mg 100 ml^{-1}) to the buffer-substrate solution appropriate to the enzyme under study. The method is most obviously applicable to the study of isoenzymes of dehydrogenases which require one or other of the coenzymes as second substrates. However, the value of the method is extended much beyond this class of enzymes to include transferases such as aspartate aminotransferases and creatine kinase[9] when suitable coupling reactions are used. In some techniques, *e.g.*, for aminotransferases, in which the coupled enzymic reactions lead to the oxidation rather than the reduction of the coenzymes, the enzyme zones appear as pale areas in a dark background.[45] The reduction of a tetrazolium salt to a formazan by reduced PMS has even been extended to the detection of zones of dipeptidase and tripeptidase activity after electrophoresis on polyacrylamide gel or cellulose acetate.[46] In this application, amino acids released by enzyme action are oxidised by L-amino-acid oxidase with simultaneous reduction of flavine adenine dinucleotide, which in turn reduces PMS.

For enzymes of relatively low specificity, the chemical nature of the substrate can be varied so that products are formed which are themselves coloured, or which can readily be converted to coloured compounds. Chromogenic substrates include *p*-nitroaniline linked by a peptide bond to various amino acids to form derivatives which are acted on by arylamidases or amino-acid transferases (*e.g.*, γ-glutamyl transferase, EC 2.3.2.2), and phosphate esters of *p*-nitrophenol which are hydrolysed by non-specific acid or alkaline phosphatases giving rise in alkaline solution to the yellow *p*-nitrophenate ion. Esters of indoxyl or halogenated indoxyls with orthophosphoric acid[47] or aliphatic carboxylic acids[48] also serve as chromogenic substrates for appropriate esterases, indoxyl or its derivatives produced by enzymic action being oxidised by air or potassium ferricyanide to indigo dyes.

The yellow colours of *p*-nitroaniline or *p*-nitrophenol are rather pale, and these compounds are also quite soluble and diffusible. A more useful series of esterase substrates is derived from various naphthols (*e.g.*, 1- or 2-naphthol, naphthol AS, or naphthol AS-MX) or naphthylamines. Although the products of the enzymic reactions are not themselves coloured, they react readily with stabilised diazonium salts to give intensely coloured, insoluble dyes.[48–51] Several alternative diazonium salts have been preferred by different investigators on the grounds of reactivity or low background staining; one of the most useful is Fast Blue B (tetrazotised *o*-di-anisidine). Coupling between naphthols and diazonium salts is more rapid at alkaline than acid pH, and 1-naphthol is generally more reactive than 2-naphthol.

The ability to vary the nature of the substrate is useful in studies of the substrate specificity of individual isoenzyme zones separated electrophoretically. A series of aliphatic esters of various naphthols has been used in this way in the characterisation of mouse-liver esterases separated by starch-gel electrophoresis,[48] while the substrate specificity of human liver and small-intestinal alkaline phosphatases was assessed after starch-gel electrophoresis[52] by precipitating inorganic phosphate released from several orthophosphate and pyrophosphate substrates as calcium phosphate, which was then converted to lead phosphate and finally to a visible precipitate of lead sulphide[53] (Fig. 2.7).

Quantitative estimates of the relative proportions of different isoenzymes can be obtained by densitometric scanning of the coloured zones obtained by the staining methods outlined above. Inherently transparent supporting media such as agar or polyacrylamide gels have an advantage in this respect, although opaque media such as starch gel or cellulose acetate can be made transparent by various treatments; alternatively, reflectance scanning may be used. Limitations which apply to densitometry of non-enzymic substances also apply to measurement of the intensities of isoenzyme zones, such as non-proportionality between the amount of dye in the zone and the observed absorbance of light. However, additional constraints enter into the quantitation of isoenzyme zones. The enzymic reaction is usually allowed to proceed for a fixed period of time before it is terminated and the zones are scanned. Not only must the amount of colour produced in each zone during this period be

Liver   Small intestine

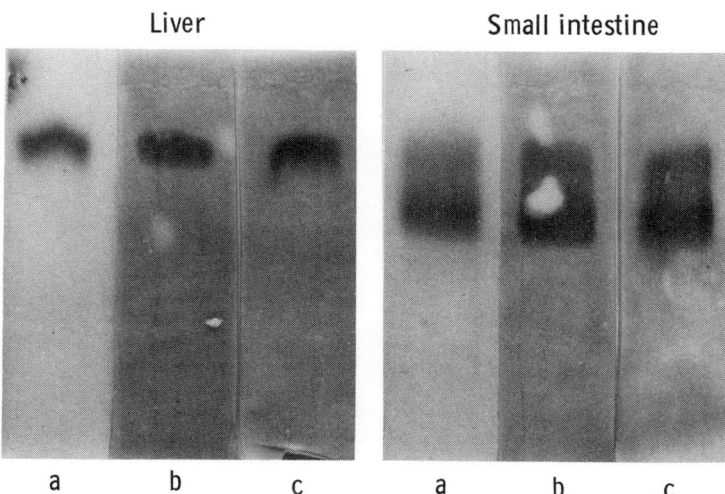

a   b   c   a   b   c

Figure 2.7   *Horizontal starch-gel electrophoresis at pH 8.6 of alkaline phosphatase preparations from human liver and small intestine, stained to show activity with three substrates: a, 1-naphthyl orthophosphate; b, adenosine diphosphate; c, uridine triphosphate. Liberation of 1-naphthol was demonstrated as in Fig. 2.3. Inorganic phosphate released from ADP and UTP was visualised by precipitating it in the gel as calcium phosphate which was subsequently converted to lead phosphate and then to lead sulphide. The anode is at the top and the origin at the bottom.* [Reproduced, with permission, from: R. H. Eaton and D. W. Moss, *Biochem. J.*, 1967, **105**, 1307]

within the measurable range, but also the conditions of reaction must be such that, in each zone, the amount of product formed is directly proportional to the amount of active enzyme in that zone. In other words, the rate of reaction in each zone must be first-order with respect to enzyme concentration, but zero-order with respect to substrate concentration throughout the reaction period. Conditions under which these relationships hold for all the enzyme fractions are difficult to select when the fractions differ significantly in kinetic properties. This is true for the isoenzymes of lactate dehydrogenase, which differ quite markedly in their Michaelis constants; therefore, substrate concentrations which represent compromises between the optima for the various isoenzymes must be used to avoid underestimation of particular isoenzymes.

Some of the sources of error inherent in the use of fixed-incubation methods with subsequent densitometry can be avoided by repeatedly scanning the electrophoretic strip at measured time intervals during the course of the enzymic reaction. If a scanner capable of measuring at

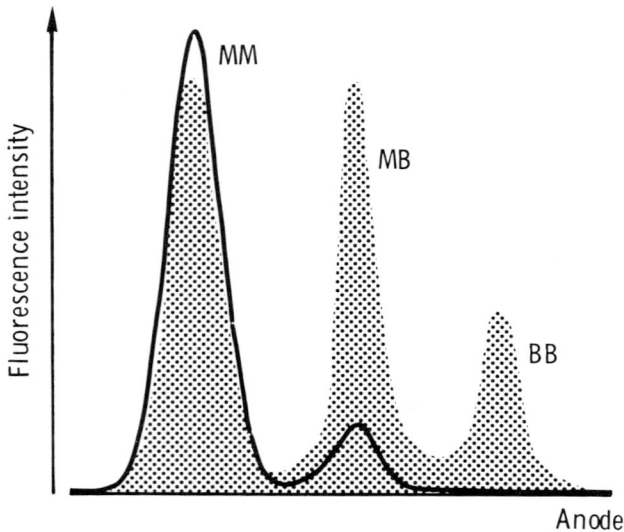

Figure 2.8 *Separation of the isoenzymes of creatine kinase in serum from a patient with recent myocardial infarction, by electrophoresis on cellulose acetate at pH 8.6. Activity was demonstrated by a series of coupled enzymic reactions, leading from ATP produced by the action of creatine kinase, through glucose-6-phosphate to reduced nicotinamide adenine dinucleotide phosphate, the fluorescence of which was recorded by scanning the strip in ultraviolet light. The shaded area shows the positions of the three isoenzymes in a control preparation separated beside the serum sample*

wavelengths near 340 nm is used, changes in oxidation of NAD or NADP can be followed without the need for further redox reactions. The successive absorbance readings at the regions of enzyme activity are used to plot progress curves of change of absorbance with time.[11] Thus, errors due to failure to follow zero-order kinetics or to lack of linearity between product concentration and absorbance can be detected. However, the technique does not eliminate possible errors due to differences between isoenzymes in such characteristics as dependence of rate of reaction on substrate concentration.

Many of the products of enzymic reactions are fluorescent when excited with light of an appropriate wavelength, so that measurement of fluorescence emission from isoenzyme zones may be an alternative to absorbance scans in the quantitation of electrophoretic patterns. Fluorescence measurements are usually one or more orders of magnitude more sensitive than absorbance measurements. However, the range of linearity between fluorescence-emission and concentration is often more limited due to quenching effects. Both NADH and NADPH are fluorescent, with wavelength maxima for excitation and emission at 340 and 435 nm, respectively, and this property has been used to increase the sensitivity of detection and measurement of the MB isoenzyme of creatine kinase after electrophoresis[20] (Fig. 2.8). Naphthols and naphthylamines have different fluorescent characteristics from their ester or amide derivatives, so that liberation of naphthols or naphthylamines by enzymic action can also be detected with excitation wavelengths of 335—345 nm and fluorescence measurement in the region of 415—455 nm, depending on the particular compound under study. This principle has been applied to the detection of zones of arylamidase[50] and alkaline phosphatase[54] after separation of human serum and tissue extracts by starch-gel electrophoresis. The most sensitive fluorigenic substrates so far described consist of esters and conjugates of umbelliferone (7-hydroxycoumarin) or 4-methylumbelliferone (4-methyl-7-hydroxy-coumarin). In alkaline solution the methyl umbelliferone anion has an excitation wavelength of 360 nm and maximum emission at 448 nm, and its high extinction coefficient and fluorescence quantum yield allow it to be detected with great sensitivity. Substrates derived from this compound have been used to locate zones of hexosaminidase after electrophoresis on starch gel or other media, as well as various other isoenzymes.[55] When conditions in the reaction mixture (*e.g.*, pH) are favourable for fluorescence, the progress of reaction can be followed by successive, timed observations with a suitable scanner.

Measurement of enzymic activity in solutions eluted from segments of the supporting medium after electrophoresis provides an alternative approach to quantitation of the relative proportions of isoenzymes.[54, 56–59] Although laborious, this method permits the choice of reaction conditions which are optimal for each enzyme fraction, as well as providing the opportunity of avoiding limitations of fixed-incubation techniques by the use of continuous-monitoring methods of enzyme assay. However, elution introduces new potential sources of error, in that recovery of the various isoenzyme fractions from the supporting medium may be low, reducing analytical sensitivity, or, more importantly, the recoveries of different isoenzymes may not be equal. Good recoveries from cellulose acetate or starch block can be obtained by passive diffusion from the supporting medium into solution; however, more vigorous treatments are required when gel media are used. Segments of starch gels can be macerated with a suitable buffer solution, or can be first frozen then thawed, a process which converts the gel to a spongy substance from which the enzyme-containing fluid can be expressed, or expelled by centrifugation. Digestion of starch gel with α-amylase has been applied to aid recovery of isoenzymes.[60] Electrophoretic elution from both starch and polyacrylamide gels has also been used.

In some methods, substances produced by isoenzyme action *in situ* in gels or other media, rather than the isoenzymes themselves, are eluted for colorimetric or fluorimetric analysis in solution. A wider range of product concentration may be measured in this way, compared with scanning procedures.[61] However, the inability to vary the reaction conditions to suit the characteristics of individual isoenzymes applies to this method, as it does to methods in which the reaction products are determined at their sites of generation.

A potential source of error which is present in all quantitative electrophoretic analysis of mixtures of multiple forms of enzymes, whatever means are used to detect and measure the separated zones, is the possibility that certain components of the mixture will be inactivated to a relatively greater extent than others during the process of separation. For example, the several zones of alkaline phosphatase in extracts of a single human tissue have similar stabilities to heat (although there are differences between tissues in this respect) and therefore presumably do not undergo differential losses of activity during electrophoresis.[56] This is not true, however, of the isoenzymes of lactate dehydrogenase, among which the more-cathodal forms 4 and 5 are significantly less stable to heat than the more-anodal isoenzymes with which they co-exist in some tissues.[62] For this enzyme, therefore, thermal inactivation of isoenzymes 4 and 5 is an important potential source of error in quantitative electrophoretic analysis.

Inactivation of isoenzymes during electrophoresis can be reduced by minimising heating as far as possible. This can be done by using short separation times on small, thin films or plates, *e.g.*, of cellulose acetate. For longer separations, the apparatus may be placed in a refrigerator or cold room. More active cooling may be necessary when gel media are used and may be achieved by surrounding the gel with a volatile organic solvent,[21] or by circulating cold water through channels close to the gel surfaces.[36, 37] Some particularly labile isoenzymes can be protected from inactivation by adding suitable reagents to the electrophoresis buffer solutions; addition of albumin (1% m/V) improves the recovery of isoenzyme 5 of lactate dehydrogenase in agar gel[10] or cellulose acetate[63] electrophoresis and dithiothreitol (1 mmol l^{-1}) that of the BB isoenzyme of creatine kinase after electrophoresis on cellulose acetate.[64]

Preparative Electrophoresis.—The high resolving power of zone electrophoresis offers a means of significantly increasing the specific activities of enzyme solutions in a single step; for example, starch-gel electrophoresis of solutions of human alkaline phosphatases from several tissues increased their specific activities by factors of from 3- to 125-fold, depending on the protein content of the original solution, yielding preparations sufficiently pure for reproducible comparisons of properties to be made.[54] These and similar observations illustrate the value of preparative electrophoresis as an aid to isoenzyme characterisation. Small-scale preparative electrophoresis can be carried out by elution of isoenzyme zones from the supporting medium after electrophoresis essentially as described for the quantitation of isoenzymes by assays of activity in solution. However, the volume of sample which can be applied to many electrophoretic media is small, while recoveries may be reduced by adsorption of isoenzymes to the support. In the experiments with alkaline phosphatase already mentioned, the combination of these factors necessitated the use of sensitive fluorimetric assay techniques to investigate the properties of the recovered enzyme fractions.[54, 56] Nevertheless, recovery of isoenzyme zones by elution from the supporting medium has been used quite extensively in isoenzyme studies.

Starch-block electrophoresis, in which the support is a paste of potato-starch grains mixed with buffer solution, can be used to separate reasonably large volumes of sample (of the order of 0.5 ml cm^{-2} of cross-sectional area of the block), while recoveries of enzyme activity after electrophoresis are usually good and can approach 100%. Isoenzyme zones are eluted from segments cut from the block by stirring the segments with buffer solution then centrifuging, or by washing them on a sintered-glass funnel. The starch-block technique was used in early studies of lactate dehydrogenase isoenzymes to determine their individual Michaelis constants and pH optima,[57] and in the investigation of multiple forms of alkaline phosphatase.[65, 66, 67] Certain synthetic materials can replace starch grains in block electrophoresis, with advantages of reduced electro-endosmotic flow and absence of interference with analysis of recovered enzyme fractions by particles present in the eluates. Various polyvinyl chloride resins have been used; perhaps the most useful of these is Pevikon C-870, a co-polymer of vinyl chloride and vinyl acetate, which has been applied to the separation of alkaline phosphatases in serum.[68] Glass powder or beads are also suitably inert supports for preparative zone electrophoresis, while other useful materials are sponge rubber or plastic foam. Rectangular pieces of these substances soaked in buffer solution are packed side by side in a plastic tray to form the electrophoresis block, with the divisions between the pieces at right angles to the direction of migration. After electrophoresis the buffer solution containing dissolved enzyme is squeezed out of each segment.[69, 70]

Continuous electrophoresis can be applied to the separation of isoenzymes on a preparative scale. In this technique, the sample mixture is supplied continuously during electrophoresis to the top of a vertically arranged supporting medium, across which a potential difference is applied. Individual protein components are thus diverted to varying degrees towards the anodal or cathodal sides of the support during their downwards migration and the separated fractions are collected in tubes placed in a row below its lower edge. Sheets of thick filter-paper usually serve as the stabilising medium, although alternatives such as glass fibre, starch grains, or sand can be employed, and continuous electrophoresis in thin liquid films without stabilisers has also been described. However, although continuous electrophoresis on filter-paper was used in some of the earliest studies of lactate dehydrogenase isoenzymes,[71] the method is technically complicated and now seems to find little application.

The great resolving power of electrophoresis in starch or polyacrylamide gels has focused considerable interest on the adaptation of this technique to the preparative-scale investigation of enzyme heterogeneity. Several designs of apparatus have been described, usually intended for use with polyacrylamide gel,[72-78] and their number and variety reflect not only the extent of the difficulties which have been encountered in evolving

---

Figure 2.9 (Top) *Construction of a preparative polyacrylamide-gel electrophoresis apparatus. A, borosilicate glass tube which forms the neck of a spherical glass electrode vessel of* 1 1 *capacity; B, central cooling tube; C, L, upper and lower rings; E, M, rubber 0-rings, upper ring being compressed by insert D; F, conical chamber used in casting the gel (shown hatched), allowing the gel to be drawn into the pores of the sintered-glass disc, G, which forms the upper boundary of the elution chamber, H, thus anchoring the gel during electrophoresis. Passage of proteins through the elution chamber during electrophoresis is prevented by the dialysis membrane, J, supported on a sintered-glass disc, K, pierced with holes (Annular area of the gel is* 10 cm²)

(Bottom) *Plan and section of the elution chamber. The stippled area is swept by the elution buffer*
[Reproduced, with permission, from: J. K. Smith and D. W. Moss, *Anal. Biochem.* (Academic Press), 1968, **25**, 500]

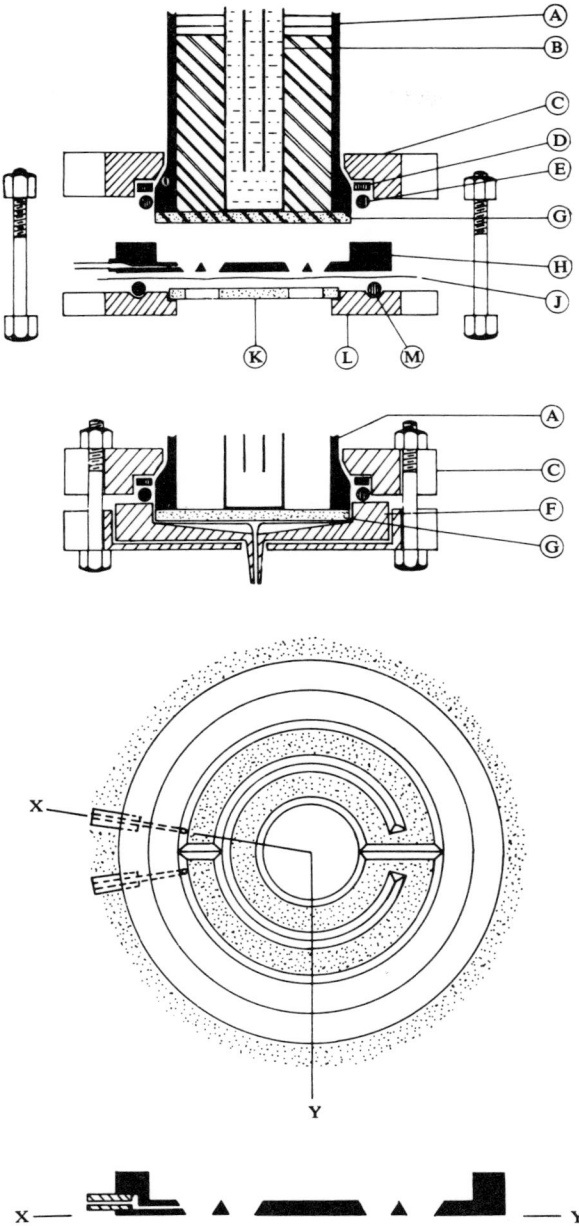

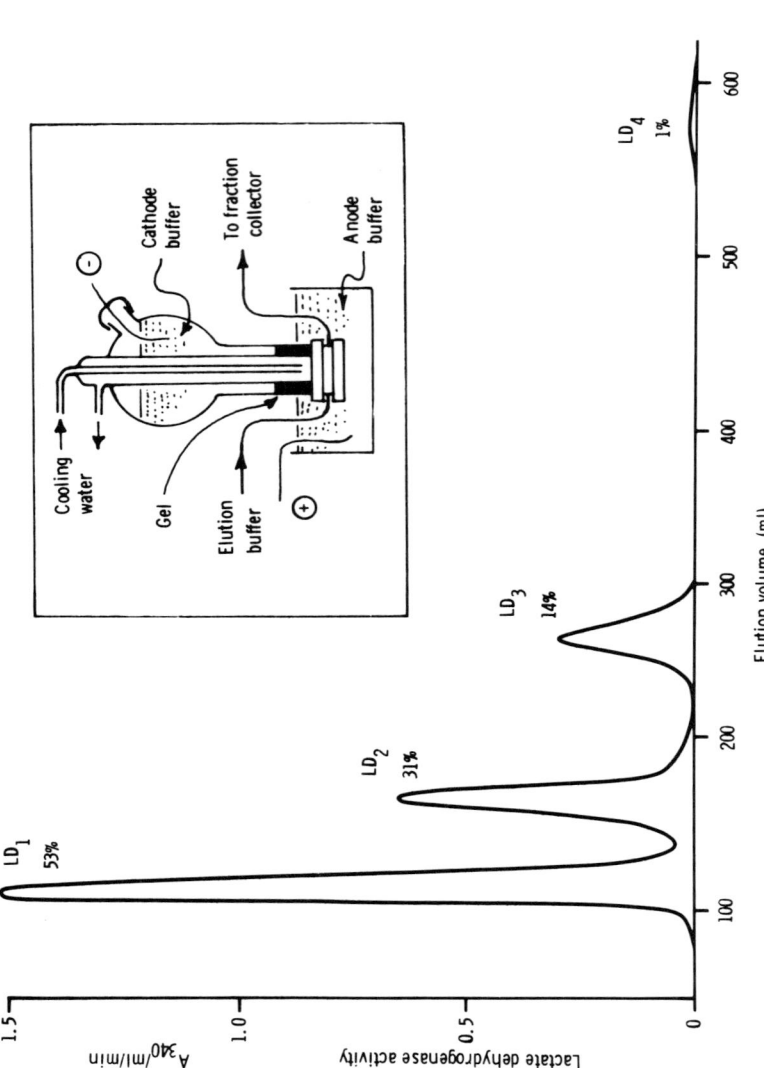

Figure 2.10 *General arrangement of the assembled preparative polyacrylamide-gel electrophoresis apparatus (inset) and pattern of elution of a mixture of lactate dehydrogenase isoenzymes. The recovery of applied activity was 99% in this experiment.*

a satisfactory system, but also the value which has been attached to this preparative technique. Certain design features have recommended themselves to several investigators, including downward migration through vertical gels, to reduce problems of sample application, and annular gels to allow both central and peripheral cooling. However, critical features in each apparatus are the arrangements made to recover quantitatively the separated protein zones as they leave the gel, while preserving the resolution achieved during separation. Elution has been effected by a continuous, transverse flow of buffer, or by intermittent sampling from a collecting chamber. The volume of the elution chamber must be small, to preserve zone sharpness. The apparatus[76] shown in Figs. 2.9 and 2.10 is able to resolve isoenzyme zones differing in mobility by 10% or more in gels 3 cm in depth, with recoveries of activity greater than 90%. However, the method is less suitable for molecules with a low net charge or high molecular weight which migrate slowly into and through the gel, causing zone broadening. Increasing the voltage gradient to offset this effect increases heat production and may inactivate the enzyme fractions.

Isoelectric Focusing

This is probably the most highly resolving of all separation techniques which make use of differences in the ionisation characteristics of protein molecules. It depends on the principle that, when a potential difference is applied to a stabilised pH gradient, components of a protein mixture will migrate electrophoretically through the gradient until they each reach the region in which the pH corresponds to their respective isoelectric points. Since a protein molecule is electrically neutral at its isoelectric pH, no further migration takes place and a series of stationary zones of proteins of different isoelectric points is formed. Continued application of the potential gradient causes the zones to become more compact, or 'focused'. Although not a new concept, isoelectric focusing has only become practicable in recent years, when the problems of maintaining a uniform, stationary pH gradient have been overcome. This is due to the introduction of synthetic carrier ampholytes consisting of polyamino-polycarboxylic acids with isoelectric points covering the range from pH 3.5 to 9.5 ('Ampholine'; LKB-Produkter AB, Bromma, Sweden). These substances migrate in an electric field to their individual isoelectric points at which they maintain a constant pH by their inherent buffering capacity. The pH gradient is stabilised against disruption by convection currents by a sucrose density gradient, or by polyacrylamide gel or other media which promote little or no electro-endosmotic flow.

   Both preparative and analytical applications of isoelectric focusing have been described. Preparative separations are usually carried out in a sucrose density gradient in water-cooled, vertical glass columns. Although higher potential differences favour the production of more discrete zones, the applied voltage is limited by local heating effects and consequent convection currents, because of non-uniform conductivity developing during focusing. When equilibrium is reached, which may require several days, the focused protein zones are recovered by emptying the column into a fraction collector. An example of the application of the technique to isoenzyme studies is the resolution of the MM-creatine kinase isoenzyme of human serum into three fractions.[79] Some of the high resolution of the focused zones is lost by diffusion once the applied potential is switched off and by mixing as the column is emptied. Several ingenious arrangements have been proposed for the isolation of separated fractions without re-mixing: one example consists of a line of v-shaped compartments, with a lid formed from a complementary series of v-shaped projections.[80] The ampholyte solution and protein sample are placed in the compartments. When the lid is closed, displacement of liquid over the divisions between the compartments forms a continuous zig-zag channel of liquid between the electrodes at the opposite ends of the apparatus. When focusing is complete, the lid is removed, breaking the liquid continuum and leaving the separated fractions within the individual compartments. Diffusion is therefore prevented. Furthermore, since the apparatus is horizontal, proteins precipitated at their isoelectric points fall to the bottom of the compartments. However, the number of fractions obtainable is limited to the number of compartments.

Extremely high resolution of protein zones on an analytical scale has been achieved by isoelectric focusing in 'Ampholine' pH gradients formed within rods or plates of polyacrylamide gel. Among the enzymes and isoenzymes to which these techniques have been applied are L-amino-acid oxidase,[81] lactate dehydrogenase,[82] and alkaline phosphatase.[83]

The very high resolving power of isoelectric focusing is not obtained without some potentially serious disadvantages in isoenzyme separations. The solubility of protein molecules is typically at its lowest at the isoelectric pH, so that precipitation may occur. Stability may also be reduced and loss of activity is not unusual in isoelectric focusing of enzymes. This may be reversible when it results from loss of a constituent metal atom. If the sample is introduced towards the extremes of the pH gradient, enzyme molecules may be exposed for some time to acid or alkaline pH conditions, or to oxidising or reducing conditions due to electrolysis at the electrodes, which can be additional causes of denaturation. The possibility of artefacts due to interaction of isoenzyme fractions with carrier ampholytes must also be recognised.

Chromatography

Partition of protein molecules between the distinct phases of a two-phase system has been for many years an established technique for the selective isolation of a particular enzyme from contaminant proteins, as a stage in enzyme purification. A well-known example is the adsorption of enzymes to calcium phosphate gel and their subsequent elution. However, the principle of continuously repeated distributions between a moving and a stationary phase, constituting chromatography, is of more recent application in enzyme fractionation. The first such technique to be suitable for this purpose, ion-exchange chromatography on substituted-cellulose ion-exchangers,[85] had a profound effect on the development of concepts of enzyme heterogeneity. This type of chromatography continues to be of great practical importance, while gel filtration and affinity chromatography have subsequently been added to the range of chromatographic techniques which are available for isoenzyme separations.

Ion-exchange Chromatography.—In the fractionation of protein mixtures by ion-exchange chromatography, the mobile phase is a solution of the protein mixture in buffer solution and the stationary phase consists of ionised groups attached covalently to an insoluble matrix of cellulose or cross-linked dextran polymers ('Sephadex'; Pharmacia AB, Uppsala, Sweden). The open structures and hydrophilic characteristics of these matrices allow access of large protein molecules to the ionised groups. The most useful of these groups for the fractionation of isoenzyme mixtures are carboxymethyl (CM), which is negatively charged in the range from approximately pH 3 to 6 and can be used to separate molecules which are cations in this pH range, and diethylaminoethyl (DEAE), which acquires a positive charge between approximately pH 6 and 9 and which therefore acts as an anion-exchanger in this pH region. Elution of protein components from the ion exchanger in order of increasing net charge at the initial pH is effected by a gradient of pH, or of increasing concentration of a counter-ion (*e.g.*, chloride in the case of an anion exchanger).

The choice of either cation- or anion-exchange chromatography depends on such factors as the isoelectric points of the components of the mixture to be fractionated and their relative stabilities on the acid or alkaline sides of their isoelectric pH values. Anion-exchange chromatography is more often used than cation-exchange in the fractionation of isoenzyme mixtures. Chromatographic fractionations are qualitatively similar to those obtained by zone electrophoresis at a similar pH value, although resolution by chromatography is generally inferior to that achieved by the best analytical electrophoresis techniques

(Fig. 2.11). However, comparatively large volumes of solution can be fractionated by ion-exchange chromatography and recoveries of enzyme activity are usually good, so that the method is particularly useful for purification of individual isoenzymes or for quantitative analysis of isoenzyme mixtures. Recovery of enzyme activity is sometimes improved by addition of detergents, reducing agents, or co-factors such as metal ions to the buffer solutions used for chromatography or for pre-treatment of the ion-exchange material.[86]

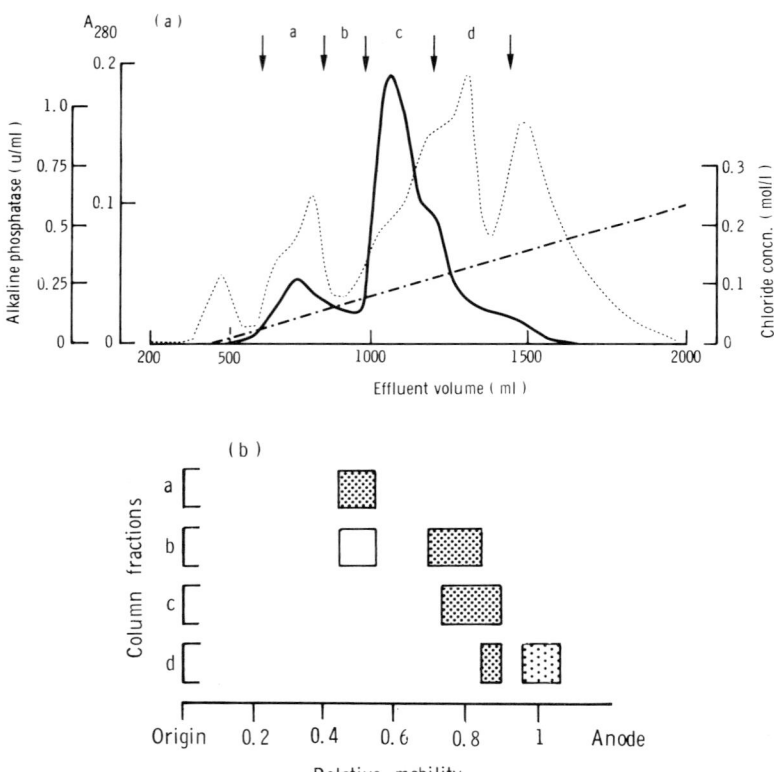

Figure 2.11   (a) *Anion-exchange chromatogram of a preparation of human intestinal alkaline phosphatase. The enzyme (solid line) was eluted from a column of* DEAE–*cellulose* (90 × 2.5 cm) *at pH* 7.7 *by an increasing concentration of chloride ions (chain-dotted line). Protein concentration, measured by absorbance at* 280 nm, *is shown by the dotted line.*
(b) *Diagram of the enzyme zones separated by horizontal starch-gel electrophoresis of fractions* a, b, c, *and* d *from the ion-exchange column.* (Modified from reference 86.)

Fractionation of mixtures of multiple forms of enzymes on ion-exchange columns with gradient or stepwise changes in buffer composition has been described for numerous enzymes, including among others alkaline phosphatase,[86,87] lactate dehydrogenase,[88] and creatine kinase.[89] Although rather more complicated to operate, gradient elution provides better resolution of molecular species with closely similar charge properties than is achieved with stepwise elution. The steepness of the gradient can be varied to produce the optimum combination of resolution and zone-width: resolution of closely similar peaks is improved by shallow gradients, whereas sharper peaks are eluted when steep gradients are used. Gradient elution also reduces the danger inherent in the stepwise technique that part of a species incompletely eluted in one step may be carried over into a succeeding eluate, causing overestimation or even mis-identification of subsequent enzyme fractions. This potential source of error is also present in batchwise separations by ion exchange, in which the enzyme sample is shaken with a quantity of ion-exchange material and the adsorbed fractions are removed by successive washes with appropriate eluants. Nevertheless, the simplicity and speed of stepwise and batchwise ion-exchange chromatography are attractive for the quantitative analysis of isoenzymes in serum. A batch technique has been described for the separation of lactate dehydrogenase in serum into anodic and cathodic isoenzyme fractions by adsorption of dialysed serum on DEAE–cellulose suspended in phosphate buffer. The anodic isoenzymes are preferentially absorbed and are removed by centrifugation, activity remaining in the supernatant being a measure of the proportion of cathodic isoenzymes.[90] Quantitation of creatine kinase isoenzymes in serum by stepwise elution from small columns (about 6 × 0.5 cm) of DEAE–cellulose or DEAE–Sephadex by successive washes with solutions of lower and higher chloride concentrations is widely used in the diagnosis of myocardial infarction.[91,92] Batch chromatography with glass-bead or resin ion-exchangers has also been used for this purpose.[93,94]

Gel Filtration.—In gel filtration, or exclusion chromatography, the two phases between which distribution of the solutes takes place are identical in composition but not in volume. Part of the liquid phase is retained within an inert gel supporting medium with pores of macromolecular dimensions. During chromatography, therefore, solute molecules can gain access in varying degrees to the part of the liquid phase contained within the gel, depending on their individual molecular dimensions. Molecules larger than the largest pores of the gel are completely excluded and emerge first from the column, followed by progressively smaller molecules.

As well as forming gels with suitable pore sizes, matrix substances for gel filtration must be hydrophilic, to allow access of protein molecules, and should not contain charged groups, to which ionic attraction of proteins might occur, nor should they show significant non-specific adsorption of proteins. Dextran gels consisting of branched polyglucose chains chemically cross-linked by glycerine ether bridges are widely used in gel filtration ('Sephadex'; Pharmacia AB, Uppsala, Sweden). These gels can be obtained with various pore sizes, the largest of which exclude proteins with molecular weights above about 800 000. Sephadex gels contain some carboxyl groups which may attract cations and repel anions. A salt solution such as sodium chloride (0.15 mol l^{-1}) or buffer solutions of similar ionic strength are therefore used as eluants with this material to reduce ionic interactions between the gel and protein molecules. Agarose, the neutral polysaccharide component of agar, can also be used to form gels suitable for exclusion chromatography ('Bio-Gel A'; Bio-Rad Laboratories, Richmond, California). The porosity of the gels is varied by changing the concentration of agarose and pore dimensions can be achieved which admit the largest macromolecules, such as nucleic acids and viruses, and even subcellular particles. The concentration of agarose in the larger-pore gels is similar to that used in agar-gel electrophoresis, *i.e.*, 1—2% m/V. The properties of readily variable pore size, inertness, and low non-specific absorption which have made polyacrylamide gels so useful in zone electrophoresis can also be exploited for gel filtration ('Bio-Gel P'; Bio-Rad Laboratories, Richmond, California). Polyacrylamide gels have useful exclusion limits up to molecular weights of 400 000. Both agarose and polyacrylamide gels may possess a small number of negatively charged groups so that, with these media also, salt solutions should be used as eluants to eliminate ion-exchange effects.

The hydrated gel particles used in gel filtration should be spherical, to ensure the close packing in chromatography columns and uniform flow of solvent which are necessary for optimum resolution. Gels with pore sizes suitable for fractionation of molecules in the range of sizes typical of enzyme molecules are soft and easily deformed by too great a hydrostatic pressure during packing or elution of the chromatography column, resulting in a serious reduction in flow rate. An upward flow of eluant is often found to give better and more constant flow rates than downward flow, probably because any fine gel particles present are less likely to block the outflow from the column. Re-cycling of samples through one or two columns of gel can be used to improve resolution, without the use of impracticably large columns.[86]

The volumes (V_e) in which protein molecules are eluted from a gel-filtration column bear an inverse, rectilinear relationship to the logarithms of the protein molecular weights, within the range in which the molecules can enter the gel. By calibrating the column with proteins of known molecular weights, this relationship can be used to determine the apparent molecular weights of proteins under study.[95] The same relationship holds for the ratio V_e/V_0, where V_0 is the void volume of the column (*i.e.*, the volume available to a molecule which is completely excluded from the gel), with the advantage that this ratio is independent of the dimensions of the column. Anomalous gel-filtration behaviour may be exhibited by proteins which are non-globular, or which undergo aggregation or dissociation under the conditions of chromatography, or for which a high degree of hydration increases the ratio between molecular size and mass beyond average values.

Gel filtration is a valuable stage in enzyme purification since recoveries of activity are usually excellent. Its specific place in isoenzyme studies is in the analysis of mixtures of multiple forms of enzymes which differ from each other in molecular size, such as a series of different aggregates of enzyme monomers with each other[96] or with non-enzymic proteins.

Affinity Chromatography.—In adsorption chromatography the solutes being separated are partitioned between the mobile liquid phase and a solid stationary phase, to which they are attached by non-covalent, relatively non-specific bonding due to van der Waals forces or hydrophobic interactions. Affinity chromatography is a special form of adsorption chromatography in which specific interactions take place between the molecules being separated and ligands attached to an insoluble matrix. The ligands may be chosen for their ability to combine with structural features of the constituents under study, *e.g.*, with certain carbohydrate residues present in glycoproteins, or for their part in interactions which depend on functional characteristics, such as the ability of enzymes specifically to combine with substrates or inhibitors, or with analogues of these substances. Although non-specific interactions between enzymes and matrices used in affinity chromatography may occur, due to ionic or hydrophobic interactions, it is the possibility of exploiting the specificity of binding sites on enzyme molecules which gives affinity chromatography its importance in isoenzyme analysis.

The matrices to which specific ligands are attached for affinity chromatography require properties similar to those needed in materials for ion-exchange chromatography or gel filtration; consequently, modified agarose, polyacrylamide, and dextran gels, as well as cellulose, have all been used for affinity chromatography. Several techniques are available for attachment of ligands to the hydroxyl groups of polysaccharide matrices including alkylation, epoxide formation with epichlorohydrin, and oxidation with periodate, but the most generally used method is reaction with cyanogen bromide.[97] Derivatives of polyacrylamide gels can be made by reacting the amide side-chains with nitrogenous compounds such as hydrazine. A wide range of ligands can be attached covalently to the primary derivatives prepared by these reactions. Some of these ligands may react with components of many enzyme or protein molecules and are therefore relatively unspecific: an example is concanavalin A, a plant lectin with a high affinity for α-D-mannosyl, α-D-glucosyl and sterically similar carbohydrate residues, and consequently for glycoproteins which contain such residues. Affinity chromatography on immobilised concanavalin A is now widely used in purification of glycoprotein enzymes, and has also been used to separate foetal and adult forms of γ-glutamyl transferase from liver.[98] Increased specificity is obtained by the use of ligands which combine with specialised binding sites of particular enzyme or isoenzyme molecules, and several derivatives or analogues of nucleotide coenzymes have been used in this way. Cibacron Blue F3G-A coupled to agarose gel ('Blue Sepharose CL-6B'; Pharmacia AB, Uppsala, Sweden) binds a wide range of enzymes with a 'dinucleotide fold' as part of their molecular structures to which nucleotide coenzymes are bound. The anodal (H_4) isoenzymes of lactate dehydrogenase from many species fail to bind to this type of adsorbent, or are readily eluted with solutions of the nucleotides nicotinamide-adenine dinucleotide (NAD) or adenosine monophosphate. In contrast, M_4 homopolymers are eluted only with reduced NAD. Heteropolymers have intermediate affinities for the ligand.[99] The homodimeric isoenzymes determined by the three human alcohol dehydrogenase loci, ADH_1, ADH_2, and ADH_3, also show differences in affinity for this ligand. In addition, the isoenzymic products of the common allelic genes at the ADH_3 locus differ markedly in their affinities for Blue Sepharose: the $\gamma^1\gamma^1$ isoenzyme is firmly adsorbed, but the $\gamma^2\gamma^2$ dimer is not absorbed, while the $\gamma^1\gamma^2$ hybrid exhibits intermediate affinity characteristics.[100] Adenosine monophosphate (AMP)–Sepharose can be used to separate the testicular isoenzyme, LD_X, from the lactate dehydrogenase isoenzymes found in other tissues, since LD_X alone is not bound to this absorbent in the presence of 1.6 mmol l^{-1} oxidised NAD.[101]

Other immobilised coenzymes or their analogues which have been applied in affinity chromatography include pyridoxal phosphate, folic acid, and biotin. Nucleotide and other coenzymes take part in the reaction cycles of many enzymes. Still greater specificity of binding is obtainable, therefore, with ligands consisting of molecules which are the substrates of enzymes of high substrate specificity, or which are highly specific inhibitors. A substrate analogue, diamino-caproyl-phenyl-trimethyl-ammonium, bound to agarose gel has been used to separate the usual form of serum cholinesterase from the atypical isoenzyme determined by a rare allelic gene.[102] The atypical isoenzyme can be eluted from the gel with sodium chloride more easily than the usual isoenzyme, as would be expected in view of the comparatively lower affinity for substrates of the atypical form of the enzyme. Mixed tetramers of the usual and atypical protomers can also be separated on this gel. Since antibodies can also be attached to gel matrices, the high degree of specificity of antigen–antibody reactions is also potentially available for the affinity chromatography of isoenzymes which differ in their antigenic properties.

The mode of attachment of these various ligands to the supporting matrix is critically important in devising effective media for affinity chromatography. An obvious requirement is that groups in the ligand which are important in the interaction between it and the enzyme should not be involved in, or masked by, linkage to the support. Steric hindrance can prevent the access of the enzyme molecules to what is usually a much smaller ligand, so that it is often necessary to insert spacer groups between the ligand and its support to achieve effective binding. A spacer arm may also enhance binding by extending the ligand beyond the region of restricted diffusion close to the matrix polymer and, by its flexibility, increase the probability of correct orientation between the ligand and the enzyme, although if the arm is too long its excessive flexibility may reduce binding.

Extension arms usually consist of a series of hydrocarbon groups, such as methylene groups, and thus may serve as centres of hydrophobic interaction with protein molecules, so that non-specific adsorption effects may mask specific attachments. The existence of non-specific adsorption effects, whether ionic or non-ionic, can impair the separative powers of an affinity chromatography process, particularly when the latter depends on relatively weak interactions. Therefore, it is important to assess the magnitude of non-specific effects when selecting conditions for affinity chromatography. Various methods for doing this can be devised, *e.g.*, by demonstrating a low affinity of an enzyme for the matrix without its attached ligand, or for the ligand in the presence of a substance which competes for the enzyme's binding site. The bound enzyme should be catalytically inactive, or have reduced catalytic activity in the case of an enzyme with multiple active centres on each molecule, if specific binding is believed to involve the active centre. Specific interactions involving attached ligands are also likely to be independent of ionic strength within the range in which enzyme–ligand binding can occur in free solution, whereas non-specific binding is usually dependent on this factor.

From the nature of affinity chromatography it is apparent that elution of bound enzymes or isoenzymes can be accomplished with a solution of the free ligand or an analogue of it. However, a change of eluant pH or ionic strength to values unfavourable to ligand-enzyme interaction is often found to be effective. Both specific and non-specific methods of elution are in use. The choice of conditions for elution depends on the nature of the interaction of the enzyme molecules with the bound ligand, and its strength compared with the affinity of the enzyme for a competing, soluble ligand used as an eluant. The extensive possibilities for variation of elution conditions in affinity chromatography further contribute to the great potential of this technique in isoenzyme separations.

An interesting recent development is the combination of affinity effects with zone electrophoresis. Incorporation of concanavalin A–Sepharose (Pharmacia AB, Uppsala) into starch gel causes retardation of isoenzymes which bind to the lectin during subsequent electrophoresis, whereas unbound isoenzymes migrate normally. Hexosaminidases A and B are separated from hexosaminidase C in this system.[103]

Other Separation Methods

Reference has already been made to the Kohlrausch regulating function in discussing the zone-sharpening effects produced by the use of discontinuous buffer systems in zone electrophoresis. This principle is employed in separations of ions in an electric field by the method of isotachophoresis. The mixture of ions being separated is introduced into the field between an electrolyte solution containing an ion with greater mobility than the ions of the sample (the leading ion), and one which contains an ion of lower mobility than the sample ions (the terminating ion). The sample ions are segregated into zones according to their relative mobilities between the leading and terminating ions. All the zones then migrate at the same speed. The inter-zone boundaries are sharpened by the use of high potential gradients; therefore, separations are carried out in capillary tubes to promote cooling and to prevent distortion of the boundaries by convection currents. Each zone contains only a single ionic species of uniform concentration throughout the zone, and the length of each zone is proportional to the amount of the corresponding component present in the original mixture. Applications of isotachophoresis in isoenzyme studies may therefore lie in the quantitative analysis of mixtures of isoenzymes.

Development of uniform, very small particles of various substances as packing materials for chromatography columns has greatly increased the resolution obtainable by various chromatographic techniques. High pressures are needed to maintain satisfactory flow-rates with these closely packed columns. The limited mechanical strengths of the hydrophilic supporting substances needed for chromatography of proteins may restrict the applicability of high-pressure chromatography to isoenzyme separations. Nevertheless, high-pressure liquid chromatography has already been applied to the quantitative analysis of mixtures of isoenzymes of creatine kinase in serum for diagnostic purposes.[104] Pyrolysis of solid enzyme samples followed by gas chromatography of the products has been shown to differentiate between isoenzymes of lactate dehydrogenase from various tissues.[105] Analytical ultracentrifugation with zonal rotors provides a means of investigating the intracellular distribution of isoenzymes, and, moreover, one which can be applied to the small samples of tissues which can safely be removed from living human organs.[106] Consequently, this technique offers the possibility of studying the patterns of activities of isoenzymes within normal cells and their alterations in disease.

References for Chapter 2

1 Moss, D.W., *Nature*, 1962, **193**, 981.
2 Cho, H.W., Meltzer, H.Y., Joung, J.I., and Goode, D., *Clin. Chim. Acta*, 1976, **73**, 257.
3 Baker, R.W.R., and Pellegrino, C., *Scand. J. Clin. Lab. Invest.*, 1954, **6**, 94.
4 Wieland, T., and Pfleiderer, G., *Biochem. Z.*, 1957, **329**, 112.
5 Jermyn, M.A., and Thomas, R., *Biochem. J.*, 1954, **56**, 631.
6 Kohn, J., *Clin. Chim. Acta*, 1957, **2**, 297.
7 Stagg, B.H., and Whyley, G.A., *Clin. Chim. Acta*, 1968, **19**, 139.
8 Korner, N.H., *J. Clin. Pathol.*, 1962, **15**, 195.
9 Rosalki, S.B., *Nature*, 1965, **207**, 414.
10 Wieme, R.J., *Clin. Chim. Acta*, 1966, **13**, 138.
11 Wieme, R.J., *Clin. Chim. Acta*, 1959, **4**, 46.
12 Blanchaer, M.C., *Clin. Chim. Acta*, 1961, **6**, 272.
13 Van der Helm, H.J., *Clin. Chim. Acta*, 1962, **7**, 124.
14 Papadopoulos, N.M., and Kintzois, J.A., *Am. J. Clin. Pathol.*, 1967, **47**, 96.
15 Haije, W.G., and De Jong, M., *Clin. Chim. Acta*, 1963, **8**, 620.
16 Hägerstrand, I., and Skude, G., *Scand. J. Clin. Lab. Invest.*, 1976, **36**, 127.
17 Boyd, J.W., *Clin. Chim. Acta*, 1962, **7**, 424.
18 Burger, A., Richterich, R., and Aebi, H., *Biochem. Z.*, 1964, **339**, 305.
19 Deul, D.H., and Van Breemen, J.F.L., *Clin. Chim. Acta*, 1964, **10**, 276.
20 Somer, H., and Konttinen, A., *Clin. Chim. Acta*, 1972, **40**, 133.
21 Wieme, R.J., 'Agar Gel Electrophoresis', Elsevier, Amsterdam, 1965.
22 Shaw, C.R., and Prasad, R., *Biochem. Genet.*, 1970, **4**, 297.
23 Harris, H., and Hopkinson, D.A., 'Handbook of Enzyme Electrophoresis in Human Genetics', North-Holland, Amsterdam, 1976.
24 Smithies, O., *Biochem. J.*, 1955, **61**, 629.
25 Smithies, O., *Biochem. J.*, 1959, **71**, 585.
26 Poulik, M.D., and Smithies, O., *Biochem. J.*, 1958, **68**, 636.
27 Moretti, J., Boussier, G., and Jayle, M.F., *Bull. Soc. Chim. Biol.*, 1957, **39**, 593.
28 Poulik, M.D., *Nature*, 1957, **180**, 1477.
29 Smithies, O., *Arch. Biochem. Biophys.*, 1962, **Suppl. 1**, 125.

30 Hunter, R.L., and Markert, C.L., *Science*, 1957, **125**, 1294.
31 Lawrence, S.H., Melnick, P.J., and Weimer, H.E., *Proc. Soc. Exp. Biol. Med.*, 1960, **105**, 572
32 Harris, H., Hopkinson, D.A., and Robson, E.B., *Nature*, 1962, **196**, 1296.
33 Fawcett, J.S., and Morris, C.J.O.R., *Separation Studies*, 1966, **1**, 9.
34 Ornstein, L., *Ann. N.Y. Acad. Sci.*, 1964, **121**, 321.
35 Davis, B.J., *Ann. N.Y. Acad. Sci.*, 1964, **121**, 404.
36 Raymond, S., *Clin. Chem.*, 1962, **8**, 455.
37 Ritchie, R.F., Harter, J.G., and Bayles, T.B., *J. Lab. Clin. Med.*, 1966, **68**, 842.
38 Akroyd, P., *Anal. Biochem.*, 1967, **19**, 399.
39 Hjerten, S., Jerstedt, S., and Tiselius, A., *Anal. Biochem.*, 1965, **11**, 219.
40 Margolis, J., and Kenrick, K.G., *Anal. Biochem.*, 1968, **25**, 347.
41 Shapiro, A.L., Vinuela, E., and Maizel, J.V., Jr., *Biochem. Biophys. Res. Commun.*, 1967, **28**, 815.
42 Weber, K., and Osborn, M., *J. Biol. Chem.*, 1969, **244**, 4406.
43 Sjovall, K., and Jergil, B., *Scand. J. Clin. Lab. Invest.*, 1966, **18**, 550.
44 Markert, C.L., and Møller, F., *Proc. Natl. Acad. Sci. U.S.A.*, 1959, **45**, 753.
45 Boyde, T.R.C., and Latner, A.L., *Biochem. J.*, 1962, **82**, 51P.
46 Sugiura, M., Ito, Y., Hirano, K., and Sairaki, S., *Anal. Biochem.*, 1977, **81**, 481.
47 Sugiura, M., and Hirano, K., *Biochem. Med.*, 1977, **17**, 222.
48 Hunter, R.L., and Burstone, M.S., *J. Histochem. Cytochem.*, 1960, **8**, 58.
49 Estborn, N., *Nature*, 1959, **184**, 1636.
50 Panveliwalla, D.K., and Moss, D.W., *Biochem. J.*, 1966, **99**, 501.
51 Orlowski, M., and Szczeklik, A., *Clin. Chim. Acta*, 1967, **15**, 387.
52 Eaton, R.H., and Moss, D.W., *Biochem. J.*, 1967, **105**, 1307.
53 Allen, J.M., and Hyncik, G., *J. Histochem. Cytochem.*, 1963, **11**, 169.
54 Moss, D.W., Campbell, D.M., Anagnostou-Kakaras, E., and King, E.J., *Biochem. J.*, 1961, **81**, 441.
55 Robinson, D., Price, R.G., and Dance, N., *Biochem. J.*, 1967, **102**, 525.
56 Moss, D.W., and King, E.J., *Biochem. J.*, 1962, **84**, 192.
57 Vesell, E.S., and Bearn, A.G., *J. Clin. Invest.*, 1961, **40**, 586.
58 Pfleiderer, G., and Wachsmuth, E.D., *Biochem. Z.*, 1961, **334**, 185.
59 Roberts, R., Henry, P.D., Witteveen, S.A.G.J., and Sobel, B.E., *Am. J. Cardiol.*, 1974, **33**, 650.
60 Plagemann, P.G.W., Gregory, K.F., and Wroblewski, F., *J. Biol. Chem.*, 1960, **235**, 2282.
61 Ogunro, E.A., Hearse, D.J., and Shillingford, J.P., *Cardiovascular Res.*, 1977, **11**, 94.
62 Plagemann, P.G.W., Gregory, K.F., and Wroblewski, F., *Biochem. Z.*, 1961, **334**, 37.
63 Rosalki, S.B., and Montgomery, A., *Clin. Chim. Acta*, 1967, **16**, 440.
64 Hall, N., and DeLuca, M., *Anal. Biochem.*, 1976, **76**, 561.
65 Keiding, N.R., *Scand. J. Clin. Lab. Invest.*, 1959, **11**, 106.
66 Keiding, N.R., *Clin. Sci.*, 1964, **26**, 291.
67 Cooke, K.B., and Zilva, J.F., *J. Clin. Pathol.*, 1961, **14**, 500.
68 Nordentoft-Jensen, B., *Clin. Sci.*, 1964, **26**, 299.

69 Davidson, H.M., *Fed. Proc., Fed. Am. Soc. Exp. Biol.*, 1957, **16**, 169.
70 Wills, E.D., *Biochem. J.*, 1958, **69**, 17P.
71 Sayre, F.W., and Hill, B.R., *Proc. Soc. Exp. Biol. Med.*, 1957, **96**, 695.
72 Racusen, D., and Calvanico, N., *Anal. Biochem.*, 1964, **7**, 62.
73 Jovin, T., Chrambach, A., and Naughton, M.A., *Anal. Biochem.*, 1964, **9**, 351.
74 Maizel, J.V., Jr., *Ann. N.Y. Acad. Sci.*, 1964, **121**, 382.
75 Gordon, A.H., and Lewis, L.N., *Anal. Biochem.*, 1967, **21**, 190.
76 Smith, J.K., and Moss, D.W., *Anal. Biochem.*, 1968, **25**, 500.
77 Brownstone, A.D., *Anal. Biochem.*, 1969, **27**, 25.
78 Hodson, A.W., and Latner, A.L., *Anal. Biochem.*, 1971, **41**, 522.
79 Wevers, R.A., Wolters, R.J., and Soons, J.B.J., *Clin. Chim. Acta*, 1977, **78**, 271.
80 Valmet, E., *Sci. Tools*, 1969, **16**, 8.
81 Hayes, M.B., and Wellner, D., *J. Biol. Chem.*, 1969, **244**, 6636.
82 Dale, G., and Latner, A.L., *Lancet*, 1968, **i**, 847.
83 Smith, I., Lightstone, P.J., and Perry, J.D., *Clin. Chim. Acta*, 1971, **35**, 59.
84 Latner, A.L., *Ann. N.Y. Acad. Sci.*, 1973, **209**, 281.
85 Sober, H.A., and Peterson, E.A., *J. Am. Chem. Soc.*, 1954, **76**, 1711.
86 Smith, J.K., Eaton, R.H., Whitby, L.G., and Moss, D.W., *Anal. Biochem.*, 1968, **23**, 84.
87 Grossberg, A.L., Harris, E.H., and Schlamowitz, M., *Arch. Biochem. Biophys.*, 1961, **93**, 267.
88 Hess, B., and Walter, S.I., *Klin. Wochenschr.*, 1960, **38**, 1080.
89 Yasmineh, W.G., and Hanson, N.Q., *Clin. Chem.*, 1975, **21**, 381.
90 Hess, B., and Walter, S.I., *Ann. N.Y. Acad. Sci.*, 1961, **94**, 890.
91 Mercer, D.W., *Clin. Chem.*, 1974, **20**, 36.
92 Nealon, D.A., and Henderson, A.R., *Clin. Chem.*, 1975, **21**, 392.
93 Henry, P.D., Roberts, R., and Sobel, B.E., *Clin. Chem.*, 1975, **21**, 844.
94 Morin, L.G., *Clin. Chem.*, 1976, **22**, 92.
95 Andrews, P., *Biochem. J.*, 1964, **91**, 222.
96 Harris, H., and Robson, E.B., *Biochim. Biophys. Acta*, 1963, **73**, 649.
97 Axen, R., Porath, J., and Ernback, S., *Nature*, 1967, **214**, 1302.
98 Köttgen, E., Lindinger, G., and Reutter, W., *Clin. Chim. Acta*, 1977, **80**, 221.
99 Nadal-Ginard, B., and Markert, C.L., in 'Isozymes, II Physiological Function', ed. Markert, C.L., Academic Press, New York, 1975, p. 45.
100 Adinolfi, A., and Hopkinson, D.A., *Ann. Hum. Genet.*, 1978, **41**, 399.
101 Kolk, A.H.J., Van Kuyk, L., and Boettcher, B., *Biochem. J.*, 1978, **173**, 767.
102 La Du, B.N., and Choi, Y.S., in 'Isozymes, II Physiological Function', ed. Markert, C.L., Academic Press, New York, 1975, p. 877.
103 Swallow, D.M., *Isozymes: Current Topics in Biological and Medical Research*, 1977, **1**, 19.
104 Kudirka, P.J., Busby, M.G., Carey, R.N., and Toren, E.C., *Clin. Chem.*, 1975, **21**, 450.
105 Danielson, N.D., Glajch, J.L., and Rogers, L.B., *J. Chromatogr. Sci.*, 1978, **16**, 455.
106 Peters, T.J., *Clin. Sci. Mol. Med.*, 1977, **53**, 505.

3

Selective Inactivation of Multiple Forms of Enzymes

The catalytic activity of an enzyme depends on the maintenance of the specific three-dimensional structure of its molecule, which is stabilised by numerous hydrogen bonds and hydrophobic interactions between the amino acids of the polypeptide chains composing the enzyme molecule. The conformations of the individual polypeptides, and consequently the number and strength of the stabilising bonds, are determined ultimately by the primary structures of the polypeptide chains themselves. It is not surprising, therefore, that enzyme molecules of different structures differ in the rates at which they are denatured (*i.e.*, at which the higher levels of molecular structure are disrupted) with an accompanying loss of catalytic activity, when exposed to agents which weaken hydrophobic and hydrogen bonds. Even the minor changes in amino-acid sequence, typically extending only to the replacement of a single amino acid by another, which result from allelic mutation are almost invariably accompanied by alterations, usually reductions, in enzyme stability compared with that of the common form of the enzyme. Similarly, isoenzymes derived from multiple gene loci frequently differ in resistance to inactivation, as do other multiple forms of enzymes of unknown origins.

Measurements of differences in rates of inactivation under controlled conditions are used extensively in distinguishing between multiple forms of enzymes, and can form the basis of methods of quantitative analysis of isoenzyme mixtures. Exposure to elevated temperatures or to concentrated solutions of urea or guanidine are most frequently chosen as the denaturing agents. Alternative procedures which may be useful include inactivation by oxidising or reducing reagents, or by acid or alkaline conditions. However, whatever method is chosen, careful control of experimental conditions is essential to ensure that the small differences in stability which may be associated with minor variations in structure between isoenzymes are detected and measured reproducibly.

Inactivation by Heat

The heat-stabilities of individual members of a family of isoenzymes may differ only slightly; on the other hand, marked differences may be observed. Both close similarities and wide differences in rates of inactivation by heat can co-exist amongst multiple forms of a single enzyme, *e.g.*, human alkaline phosphatase. Placental alkaline phosphatase is completely stable to heating at temperatures up to 70 °C, whereas the enzyme prepared from bone loses half its activity in less than 10 min at 55 °C at pH 7.[1,2] Although by comparison the difference in heat-stability between alkaline phosphatases from bone and liver is small, it constitutes one of the few differences in properties by which these tissue-specific forms of the enzyme can be distinguished (Fig. 3.1).

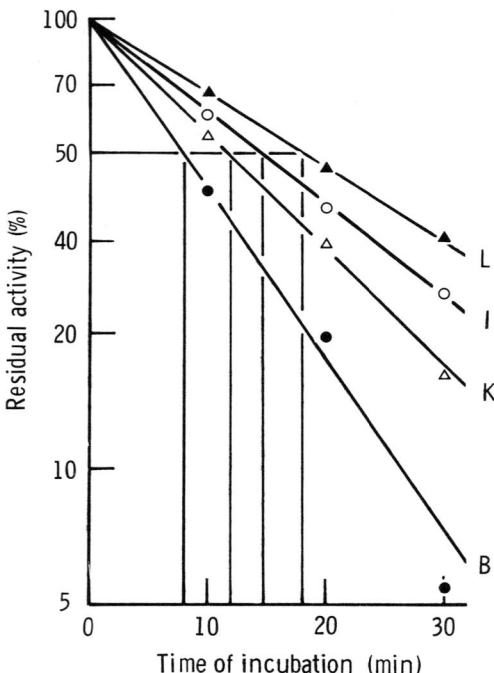

Figure 3.1 *Inactivation of alkaline phosphatases, partially purified from several human tissues by starch-gel electrophoresis, during incubation at 55 °C and pH 7. B, bone; K, kidney; I, small intestine; L, liver. The respective half-inactivation times under these conditions are approximately 8, 12, 15, and 18 min. The scale of the ordinate is logarithmic.* (Adapted from: D. W. Moss and E. J. King, *Biochem. J.*, 1962, **84**, 192)

The minor zones of alkaline phosphatase activity seen after starch-gel electrophoresis of extracts of human tissues (Fig. 1.6) have heat-stability characteristics similar to those of the main zone of activity in the same tissue extract, providing evidence that, in the case of this enzyme, the minor electrophoretic zones represent aggregates or complexes of the main enzyme component and not additional isoenzymes.[1] In contrast,

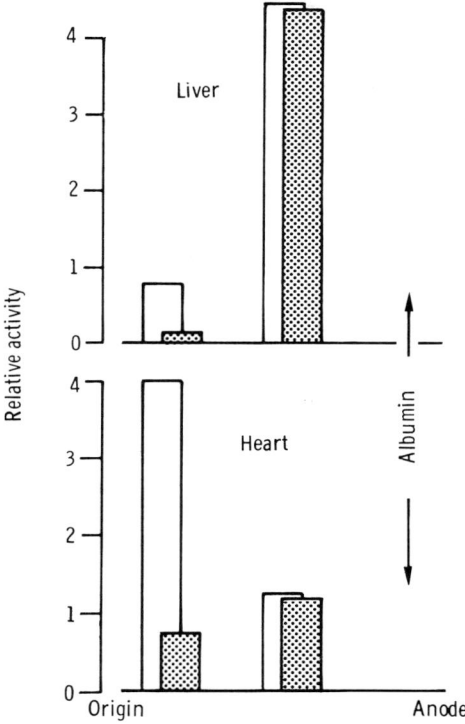

Figure 3.2 *Effect of pre-incubation for 4 h at 37 °C of extracts initially containing equal activities of human liver and heart NADP-dependent isocitrate dehydrogenase, respectively, on the relative proportions of fast and slow isoenzymes separated by horizontal starch-gel electrophoresis at pH 6.2. The open and shaded bars show relative activities in control and heated samples, respectively. The positions of the ends of the zones were located by placing strips of filter-paper soaked in the reaction mixture along the edges of the gels and observing the appearance of the blue fluorescence of NADPH, produced by enzyme action, in ultraviolet light. Solutions of the respective isoenzymes were expressed from the segments of gel thus identified after freezing and thawing, for measurement of enzyme activity.* (Based on data of D. M. Campbell and D. W. Moss, reference 4)

when electrophoretically distinct enzyme zones present in an extract of a single tissue result from quantitative differences in the expression of multiple gene loci or alleles in various cells, or in various subcellular regions such as mitochondria or cytoplasm, the isoenzymes are frequently found to differ in stability, while individual isoenzymes resemble analogous isoenzymes from other tissues in this respect. Examples of such intra-tissue differences and inter-tissue homologies are numerous.

Differences in heat-stability between mitochondrial and cytoplasmic isoenzymes have been observed for aspartate aminotransferase[3] and NADP-dependent isocitrate dehydrogenase (EC 1.1.1.42; Fig. 3.2)[4,5] as well as for other mitochondrial and extra-mitochondrial isoenzymes. Isoenzymes with different stabilities to heat determined by multiple gene loci include the A and B forms of human hexosaminidase[6] and the MM and BB isoenzymes of creatine kinase from several species,[7] as well as the well-studied isoenzymes of lactate dehydrogenase.[8] When hybrid isoenzymes consisting of unlike subunits can be formed, either *in vivo* or *in vitro*, they usually show stability properties intermediate between

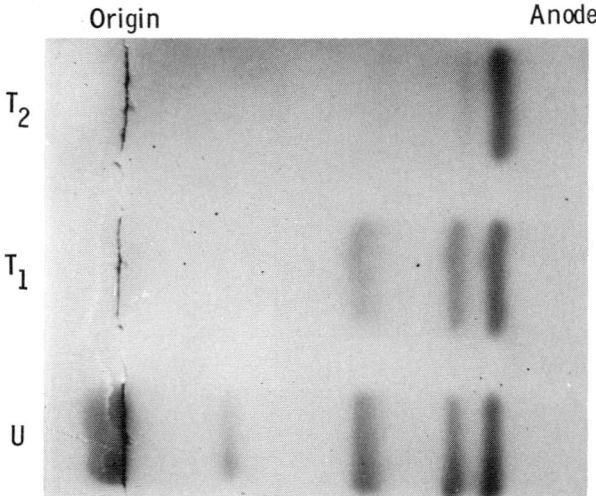

Figure 3.3   *Selective inactivation of isoenzymes by heating. A mixture of isoenzymes of lactate dehydrogenase from canine liver and heart was separated on horizontal starch-gel electrophoresis at pH 8.6 and stained for activity as in Fig. 1.8. Unheated sample (U); after heating for 30 min at 57 °C (T₁); after heating for 30 min at 65 °C (T₂).* [Reproduced, with permission, from: D. W. Moss, 'Scientific Basis of Medicine, Annual Reviews, (British Postgraduate Medical Federation, Athlone Press), 1964, p. 334]

those of the respective homopolymers; *e.g.*, the isoenzymes of lactate dehydrogenase exhibit a progressive reduction in heat-stability (Fig. 3.3), from the most-stable H_4 tetramer (LD_1) through the mixed tetramers H_3M (LD_2), H_2M_2 (LD_3), and HM_3 (LD_4) to the least-stable M_4 tetramer (LD_5). This observation is somewhat surprising, since it might be expected that denaturation of a single, less-stable M-subunit would result in the complete inactivation of any tetramer in which it occurred; *i.e.*, that the heteropolymers would resemble the M_4 homopolymer in stability. The explanation may lie in partial stabilisation of heat-sensitive protomers when associated with more stable subunits, or in the selective loss of more-labile protomers followed by a re-association of surviving subunits to form active tetramers.

The greater lability of some isoenzymes compared with others has been suggested as an explanation for their differential rates of disappearance from the circulation following elevation of enzyme activities as a result of disease. For example, a half-life of approximately 2 days has been estimated for bone alkaline phosphatase in human plasma,[9] compared with about 7 days for the more stable placental isoenzyme,[10] and radioactively labelled isoenzyme 1 of lactate dehydrogenase disappears more slowly from the blood of rabbits after injection than isoenzyme 5.[11] Plasma isocitrate dehydrogenase activity in man increases in hepatocellular damage, but not following myocardial infarction, and liver contains a greater proportion of the more stable, electrophoretically more-anodal isoenzyme than does heart.[4] However, factors other than relative isoenzyme stability probably contribute to these different responses.

Analysis of Isoenzyme Mixtures by Heat Inactivation.—Temperature coefficients for the inactivation of enzyme molecules by heat are high; *e.g.*, for human liver and bone alkaline phosphatases in serum, values of Q_{10} are 40 and 44, respectively.[12] Therefore, slight variations in temperature between measurements can have effects on rates of inactivation which are sufficiently great to obscure differences between isoenzymes with closely similar stability characteristics. Precise control of temperature during inactivation is facilitated by the use of a large-volume, well-stirred water bath controlled by a mercury-in-glass contact thermometer, and by introducing small volumes of the enzyme samples rapidly into small, thin-walled glass tubes previously placed in the bath and allowed to reach temperature equilibrium (Dreyer tubes are used in the author's laboratory). The effects of between-run variations in accuracy and precision of incubation temperature are avoided when different isoenzyme samples are compared simultaneously, but become highly significant in the day-to-day comparison of serum specimens for diagnostic purposes, or when comparing results with published values.

Rates of inactivation of isoenzymes by heat are also generally affected by changes in other factors, notably pH, protein concentration, and the concentrations of substrates and co-factors. The heat-stabilities of isoenzymes of lactate dehydrogenase and alkaline phosphatase are both pH-dependent.[13,14,15] The effects of pH on stability can be controlled by buffer solutions in the examination of tissue-extracts or purified isoenzymes. However, a considerable dilution with buffer solutions is needed to overcome the strong buffering capacity of serum samples and this adversely affects the accuracy of assays of residual enzyme activity. Measurements of the heat-stabilities of isoenzymes in serum are usually made without addition of buffer solutions, therefore, although variations in pH undoubtedly contribute to the scatter of values of heat-stability which is observed between serum samples in which a single type of isoenzyme is predominant.[14] Although increases in protein concentration generally have a stabilising effect on enzyme solutions, the BB dimer of human creatine kinase has been shown to be less stable when incubated

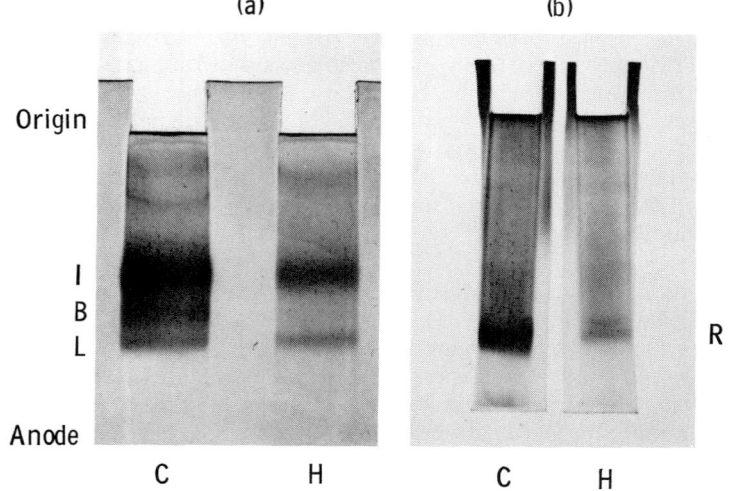

Figure 3.4   *Selective inactivation of tissue-specific forms of alkaline phosphatase by heating. Serum containing the isoenzymes was separated by vertical polyacrylamide-gel electrophoresis and stained for activity as in Fig. 2.6. (a) Control sample (C) and sample (H) heated at 56 °C for 10 min before electrophoresis, showing the relatively greater inactivation of the bone fraction (B) compared with the liver (L) and small intestine (I) enzymes. (b) Separation of unheated sample (C) compared with the pattern (H) obtained with the same sample in a segment of gel heated for 60 min at 65 °C in a sealed tube after electrophoresis but before staining for enzyme activity. Activity contributed by a carcinoplacental isoenzyme (R) survives this treatment*

at 37 °C in solutions containing human or bovine serum albumin.[16] The increased stability of enzymes in the presence of their substrates is well known and in some cases a differential effect on isoenzymes is found. For example, the heat-stability of the LD$_5$ isoenzyme is increased in the presence of NADH,[13, 17] and this substance has been incorporated in a heat-stability method for estimating the proportions of lactate dehydrogenase isoenzymes in serum.[17]

Electrophoretic discrimination of enzyme zones may be improved by comparing the patterns obtained with heated and unheated samples (Figs. 3.3, 3.4). Alternatively, selective inactivation of isoenzyme zones can be achieved *in situ* after electrophoretic separation in supporting media such as starch or polyacrylamide gels, by incubating the gels in sealed containers (Fig. 3.4) or by placing them between heated metal plates[18] before carrying out the appropriate reaction to demonstrate residual enzyme activity.

The shorter analysis times required for selective inactivation methods compared with separative procedures such as electrophoresis or chromatography have aroused considerable interest in the former approach to isoenzyme analysis in serum as an aid to clinical diagnosis. Selective inactivation methods are particularly appropriate when separation between the isoenzymes of diagnostic importance is not clear-cut, as is the case for the liver and bone forms of alkaline phosphatase, and can be adapted to automated methods of enzyme analysis. However, some of the many variables which can affect the reliability and reproducibility of heat-inactivation measurements have already been mentioned, while the need to measure low residual activities calls for methods of assay of enzyme activity of adequate sensitivity. These considerations have not always been given due weight in devising or applying heat-inactivation methods for clinical use. Nevertheless, such methods, in which residual activity is measured after heating serum samples for a fixed period, are widely employed to assess the relative proportions of the different isoenzymes of lactate dehydrogenase[17, 19-23] or alkaline phosphatase,[24-28] and measurements of the lability of amylase (EC 3.2.1.1) in serum at 65 °C have been used as an aid to the investigation of pancreatic diseases.[29, 30] Results are usually expressed and interpreted as the percentage of the enzymic activity of the unheated specimen remaining after heating; *e.g.*, less than 20% residual alkaline phosphatase activity after 10 min incubation at 56 °C suggests that the specimen contains predominantly bone phosphatase, while after 60 min at 60 °C, less than 30% residual lactate dehydrogenase activity indicates a shift towards the less-anodic isoenzymes and more than 60% a relative increase in the more anodic forms of this enzyme.

A disadvantage of these relatively simple methods is their insensitivity to small changes in isoenzyme composition, because of the wide range of heat-stabilities observed with normal specimens. Although a small increase in the proportion of, say, a heat-labile isoenzyme will result in an abnormally low heat-stability in a serum in which the value is already towards the lower end of the normal range, a considerably greater increase will be needed if the heat-stability is initially at the higher end of the normal range. Consequently, total activity of the enzyme in question typically must be more than twice the upper limit of normal before a clear segregation of a series of samples into high or low heat-stability categories is observed.[14]

A further complication is introduced by the presence of multiple isoenzyme components, some with closely similar heat-stabilities,

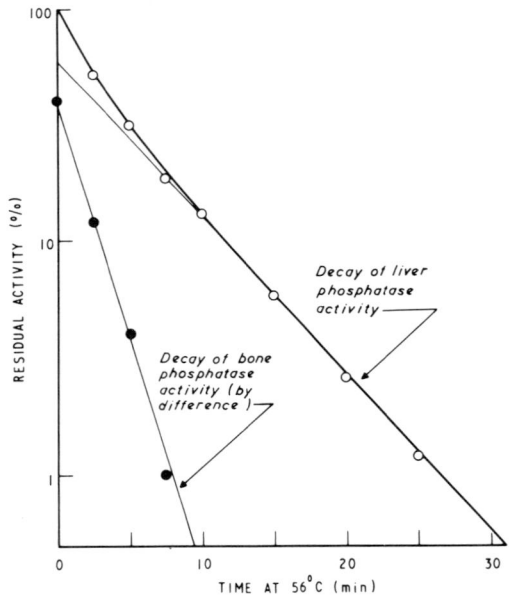

Figure 3.5   *Resolution of the curve of decline of total alkaline phosphatase activity (open symbols) with time of incubation at 56 °C into two linear components, each representing the decline in activity of one of two isoenzymes of different stabilities. The linear fall in activity between 15 and 30 min reflects the decay of the more stable liver isoenzyme, the original activity of which is obtained by producing this line to its intercept on the ordinate. The decay-curve of the less-stable bone isoenzyme (solid symbols) is deduced by subtracting the change in activity due to inactivation of liver phosphatase from the original curve. The scale of the ordinate is logarithmic.* [Reproduced, with permission, from: D. W. Moss, *Lab* (Editrice Kurtis), 1974, **1**, 363]

changes in the proportions of which may be undetectable in a single-period heating test. A separate incubation at a different temperature can detect an isoenzyme of distinctive heat-stability characteristics; *e.g.*, placental or carcinoplacental (Regan) isoenzymes of alkaline phosphatase are the only forms of this enzyme which survive heating at 65 °C for 30 min or more. Incubation for 30 min at each of two temperatures, 57 and 65 °C, has been used to subdivide lactate dehydrogenase isoenzymes in serum into the most labile (inactivated at 57 °C), the most stable (surviving incubation at 65 °C), and those of intermediate stability, represented by the difference between residual activities at 65 and 57 °C.[17]

Parallel measurements of heat-stability and of some other distinctive property can also be used to improve the analysis of multicomponent mixtures. In the case of serum alkaline phosphatase, inactivation by heat has been combined with measurements of the activity in the presence of L-phenylalanine, an uncompetitive inhibitor of the intestinal isoenzyme, to correct for the effects of the latter isoenzyme on estimates of liver phosphatase, since these two isoenzymes have rather similar heat-stabilities.[31,32]

Quantitative Analysis by Heat-inactivation Methods.—Semi-logarithmic plots of the residual catalytic activity of a mixture of isoenzymes as a function of the duration of incubation at a fixed inactivating temperature can be resolved into their component rectilinear portions, each representing the decay of a particular isoenzyme. The original activity of each isoenzyme is then derived from its corresponding decay curve. The method is most useful for two-component mixtures, *e.g.*, of the liver and bone isoenzymes of alkaline phosphatase in serum (Fig. 3.5), for which other quantitative techniques such as densitometric scanning of electrophoretic patterns are unsatisfactory.[12] Analysis of curves representing the progress of heat-inactivation does not require the assumption that heat-stabilities of specific isoenzymes are quantitatively identical in each serum specimen, provided that the qualitative differences in stabilities between isoenzymes are maintained. Also, slight variations in incubation temperature from one determination to another do not affect the comparability of results, though constancy of temperature during each estimation is obviously essential.

Construction of heat-inactivation curves by multiple estimations of residual activity is impracticable for routine clinical analysis of serum specimens. A method has been described by which residual alkaline phosphatase activity during incubation of buffered serum at 60 °C is recorded automatically with the Technicon AutoAnalyzer, but subsequent manual conversion of the data to semi-logarithmic plots of residual activity as a function of duration of incubation is required.[33] The

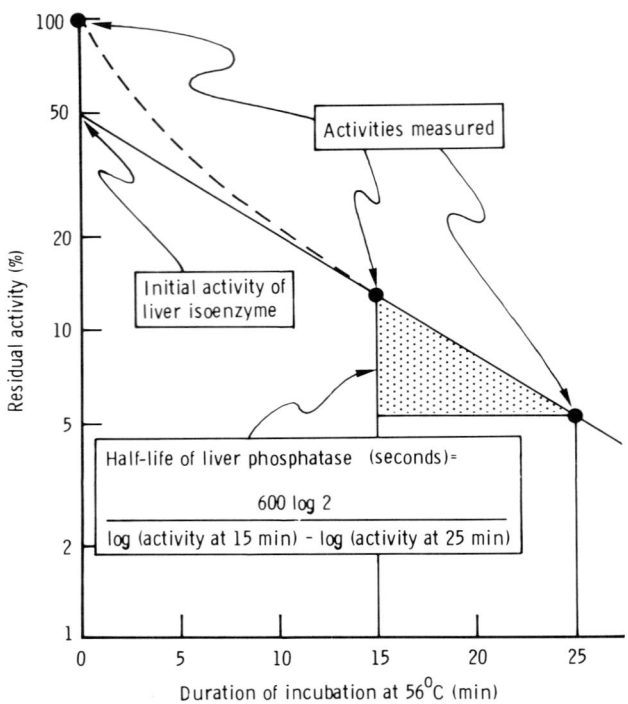

Figure 3.6 *Principle of a simplified method of resolving the decay of total activity of alkaline phosphatase into two components, so as to allow the analysis of two-isoenzyme mixtures on the basis of only three measurements of activity, according to reference 34*

principles of a method for the quantitative analysis of mixtures of liver and bone alkaline phosphatases in serum,[34] based on measurements of initial activity and activities remaining after incubation for periods of 15 and 25 min at 56 °C, are shown diagrammatically in Fig. 3.6. This simplified procedure, which retains the advantage that the results are independent of between-sample variations in isoenzyme heat-stabilities possessed by complete curves, assumes that the portion of the decay curve between 15 and 25 min essentially corresponds to inactivation of liver alkaline phosphatase alone. The validity of this assumption under various circumstances and possible sources of error, as well as analytical details of the method and some clinical applications, have been discussed extensively.[34,35] Coefficients of variation of estimates of liver and bone alkaline phosphatases in serum are of the order of $\pm 6\%$ and $\pm 10\%$, respectively, for elevated activities of the two isoenzymes.

3. *Selective Inactivation of Multiple Forms of Enzymes*

A mean half-life of liver alkaline phosphatase at 56 °C of 452 s ± 55 s.d. was found by analysis of 110 normal sera and 461 sera from patients with various diseases by the three-point method.[35] This value is close to that of 456 s ± 90 s.d. determined from full heat-inactivation curves,[12] evidence of the validity of the simplified procedure. Although the three-point method is intended for the analysis of sera in which total alkaline phosphatase activity is contributed entirely or largely by the liver and bone variants of the enzyme, values for the apparent half-life of liver phosphatase more than 2 standard deviations above the mean may result from the presence in the serum of placental or placental-like (Regan) isoenzymes.[35] A significant increase in apparent half-life can occur when

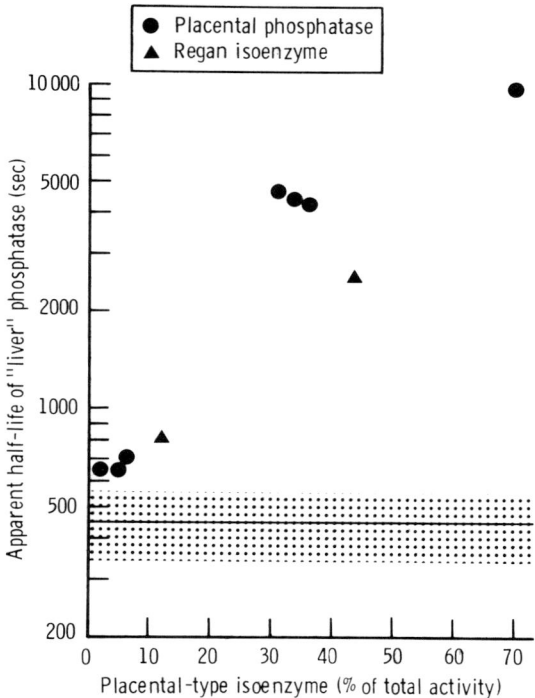

Figure 3.7 *Effect of the presence of placental or carcinoplacental (Regan) isoenzymes of alkaline phosphatase in serum on the apparent half-life of liver alkaline phosphatase during heat inactivation at 56 °C by the method outlined in Fig. 3.6. The placental and Regan isoenzymes are completely stable at this temperature so that, even in small proportions, they raise the apparent half-life beyond the range found for the liver isoenzyme. (The shaded zone indicates the mean ±2 s.d. for the half-life of liver phosphatase)*

as little as 2% of the total activity is due to these very heat-stable isoenzymes (Fig. 3.7). However, the presence of a considerable proportion of intestinal alkaline phosphatase has a relatively small lowering effect on the apparent half-life of liver phosphatase, so that the activity of the latter isoenzyme may be overestimated in these circumstances. As with all non-separative methods of isoenzyme analysis, therefore, identification of the isoenzyme components by a qualitative, separative procedure such as zone electrophoresis is recommended wherever possible.

Inactivation by Urea and Related Compounds

Concentrated solutions of urea or derivatives of urea such as guanidine and methylurea have a profound disruptive effect on higher levels of protein structure. Therefore, differential actions of these compounds on the structures and catalytic activities of multiple forms of enzymes are much used in isoenzyme analysis and characterisation. The reagents may have several effects on enzyme molecules, although these are not always distinguished in reports of their use. Exposure to high concentrations of

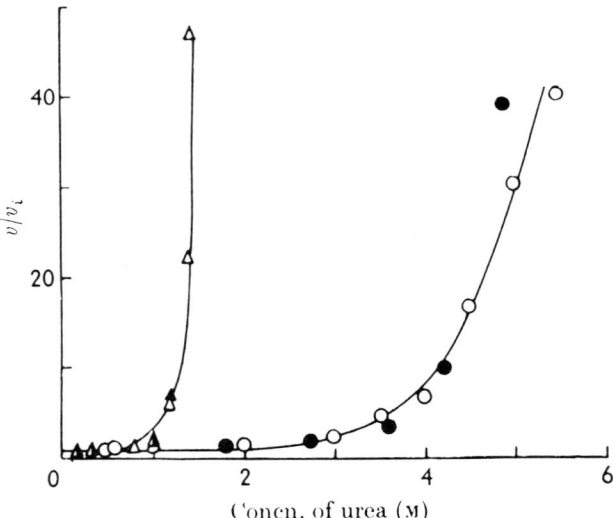

Figure 3.8(a)   *Effect of exposure to increasing concentrations of urea on the lactate dehydrogenase activities of human and ox heart (open and solid circles), and of human liver and rabbit muscle (open and solid triangles).* [Reproduced, with permission, from: W. A. Withycombe, D. T. Plummer, and J. H. Wilkinson, *Biochem. J.*, 1965, **94**, 384]

urea (6—12 moles l^{-1}) or rather lower concentrations (of the order of 5 moles l^{-1}) of guanidine may cause dissociation of polymeric isoenzymes into their component monomers, *i.e.*, quaternary structure is disturbed. This technique has been used to investigate the subunit structures of several families of isoenzymes, including among others those of lactate dehydrogenase[36] and creatine kinase.[7] In some cases, re-combination of monomers to form active enzyme molecules occurs when the urea concentration is reduced, *e.g.*, by dialysis. On the other hand, recovery of catalytic activity may be incomplete or absent, and in these instances derangement of the tertiary structures of the monomers has presumably

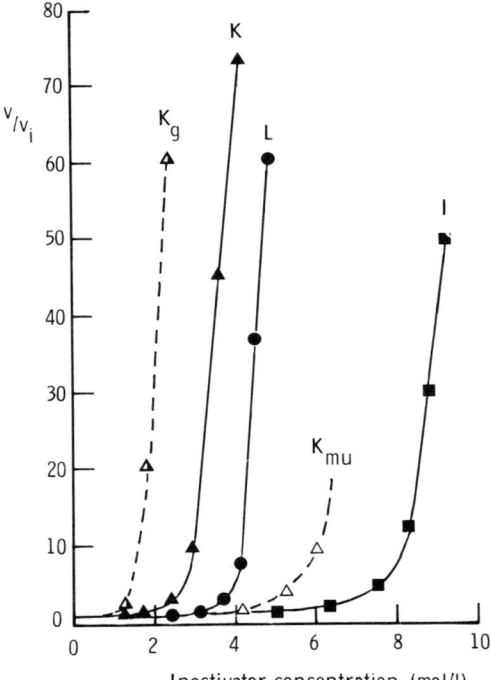

Inactivator concentration (mol/l)

Figure 3.8(b) *Effect of pre-incubation for 15 min at 37 °C with urea or analogues of urea on the activities of isoenzymes of human alkaline phosphatase. Solid triangles, circles, and squares refer to treatment with urea of the kidney (K), liver (L), and small-intestinal (I) phosphatases, respectively. Open and half-filled triangles respectively show the effects of methylurea (mu) and guanidine (g) on the kidney isoenzyme. (Adapted from reference 39)*

In both (a) and (b) the ordinate is the ratio of activity (v) in the absence of inactivator to the activity (v_i) of the partially inactivated enzymes

also occurred. In addition, reversible, inhibitory effects of low concentrations of urea have been used in distinguishing between isoenzymes. However, these seem to be related to differences in catalytic properties rather than in resistance to denaturation, and the differential effects of urea and related compounds on higher levels of molecular structure are more useful in isoenzyme analysis.[7,37–40] Plots of residual activity after treatment with urea as a function of urea concentration typically show a marked discontinuity, corresponding presumably to the concentration at which unfolding of the protein molecule takes place (Fig. 3.8).

Inactivation of isoenzymes by concentrated solutions of urea and its derivatives is markedly dependent on temperature and on time of exposure to the reagents. Other variables, such as pH and the presence and concentration of substrates, also influence rates of denaturation; consequently, all these factors must be controlled to ensure reproducible results. Cyanate, itself an inhibitor of some enzymes, forms readily in solutions of urea, which should therefore be freshly prepared.

As with denaturation by heat, selective inactivation in the presence of urea has been made the basis of quantitative analysis of mixtures of isoenzymes of lactate dehydrogenase or alkaline phosphatase in serum.[41,42] The principles of such analytical methods are similar to those used in heat-inactivation, *i.e.*, measurement of residual activity after a period of exposure to the denaturing agent, and the results of the two approaches are qualitatively similar, in that those isoenzymes which are more heat-resistant are also less readily inactivated by urea. However, the correlation is not complete, suggesting that the modes of action of the two denaturing agencies are not identical.[39,42]

Because inactivation by urea can be carried out at the same temperature at which initial and residual enzymic activity is measured, urea-inactivation methods can be incorporated into fully automated systems of analysis more easily then is the case for inactivation by elevated temperatures. Automated analysis has the advantage that, compared with manual methods, the conditions of exposure to the denaturing agent which have such a pronounced effect on the extent of inactivation can be standardised more completely for each sample. Several automated methods of isoenzyme analysis which make use of selective inactivation by urea have been described, particularly for determination of alkaline phosphatase isoenzymes in serum.[43,44,45] The scope of these methods is increased by the inclusion of L-phenylalanine to provide an estimate of the contribution of intestinal alkaline phosphatase to the total activity. As with single-period heat-inactivation techniques, however, the methods based on denaturation by urea assume that the stability properties of specific isoenzymes are quantitatively identical in each serum.

Other Selective Inactivation Procedures

Many enzymes are metalloproteins and are therefore inactivated by metal-chelating agents, such as EDTA, which remove the constituent metal atoms. As with urea, the irreversible, time-dependent inactivation of metallo-enzymes by EDTA should be distinguished from reversible inhibition, due in the case of EDTA to binding of soluble metal ions which may be activators of the catalytic process. These effects have been studied for alkaline phosphatase isoenzymes, which are zinc-containing proteins, and differences between human bone, intestinal and placental phosphatases exposed to EDTA have been demonstrated.[46] Artificial isoenzymes of bacterial alkaline phosphatase have been produced by substitution of other metal atoms for the zinc atoms of the native enzyme form. The native and modified isoenzymes have different catalytic properties.[47, 48]

Several other agents with effects on the properties of proteins in general have been shown selectively to affect the activities of particular isoenzymes under certain circumstances. Some of these, such as alcohols or other organic solvents which are protein precipitants, feature in early

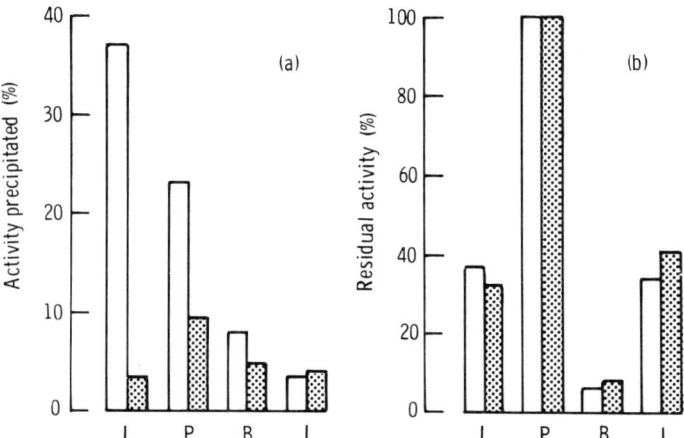

Figure 3.9 *Effect of the removal of terminal sialic acid residues by digestion with neuraminidase on the solubility* (a) *of human liver* (L), *placental* (P), *bone* (B), *and small-intestinal* (I) *alkaline phosphatases in acid ethanol. The properties of treated samples are shown as shaded bars and of control samples by open bars. Intestinal phosphatase contains no terminal sialic acid residues and its solubility is unaltered by this treatment. The stabilities of all three isoenzymes at 56 °C* (b) *are not significantly affected by the action of neuraminidase.* [Reproduced, with permission, from: R. C. Samuelson and D. W. Moss, *Clin. Chim. Acta* (Elsevier), 1978, **83**, 167]

reports of attempts to demonstrate organ-specific characteristics of enzymes. Prostatic acid phosphatase is more susceptible to inactivation by 40% V/V ethanol than the enzyme from some other tissues,[49, 50] while formaldehyde strongly inhibits erythrocyte acid phosphatase but not the prostatic isoenzyme.[51] Ethanol (20% V/V) in acid solution precipitates a greater proportion of liver or placental alkaline phosphatase than of the bone or small-intestinal isoenzymes from solution, *e.g.*, in serum.[52, 53, 54] This differential effect seems to be related to the contribution of *N*-acetyl neuraminic acid residues to the properties of the enzyme molecules, since it is largely abolished by previous incubation of the isoenzymes with neuraminidase[54] (Fig. 3.9). Exposure to extremes of pH causes denaturation of proteins. Comparison of rates of inactivation of the liver and small-intestinal isoenzymes of human alkaline phosphatase at pH 2—4 at 0 °C shows the latter isoenzyme to be more resistant to this treatment.[55]

Because of their differences in primary structure, true isoenzymes (*i.e.*, genetically distinct enzyme forms) may differ in their resistance to attack by proteolytic enzymes. Artefactual variants of some enzymes may result from proteolytic action during purification,[56] but similar changes may take place *in vivo*. Rabbit liver contains two fructose-1,6-diphos-phatases (EC 3.1.3.11) which differ in pH optima; that with the more-alkaline pH optimum can be produced by the action of papain on the enzyme which has its optimum pH close to neutrality. However, the latter enzyme can also be differentiated into two forms, one containing a tryptophan residue and present in fed animals, the other lacking both this residue and a small peptide, and found in liver in the fasting state. Although these two forms have similar catalytic properties, the 'fasted' enzyme is more susceptible to proteolysis by subtilisin.[57]

References for Chapter 3

1 Moss, D.W., and King, E.J., *Biochem. J.*, 1962, **84**, 192.
2 Neale, F.C., Clubb, J.S., Hotchkis, D., and Posen, S., *J. Clin. Path.*, 1965, **18**, 359.
3 Wada, H., and Morino, Y., *Vitam. Horm.* (*N.Y.*), 1964, **22**, 411.
4 Campbell, D.M., and Moss, D.W., *Proc. Assoc. Clin. Biochem.*, 1962, **2**, 10.
5 Islam, M., Bell, J.L., and Baron, D.N., *Biochem. J.*, 1972, **129**, 1003.
6 Robinson, D., and Stirling, J.L., *Biochem. J.*, 1968, **107**, 321.
7 Dawson, D.M., Eppenberger, H.M., and Kaplan, N.O., *J. Biol. Chem.*, 1967, **242**, 210.
8 Plagemann, P.G.W., Gregory, K.F., and Wroblewski, F., *Biochem. Z.*, 1961, **334**, 37.
9 Walton, R.J., Preston, C.J., Russell, R.G.G., and Kanis, J.A., *Clin. Chim. Acta*, 1975, **63**, 227.
10 Clubb, J.S., Neale, F.C., and Posen, S., *J. Lab. Clin. Med.*, 1965, **66**, 493.
11 Wilkinson, J.H., and Qureshi, A.R., *Clin. Chem.*, 1976, **22**, 1269.

12 Whitby, L.G., and Moss, D.W., *Clin. Chim. Acta*, 1975, **59**, 361.
13 Vesell, E.S., and Yielding, K.L., *Ann. N.Y. Acad. Sci.*, 1968, **151**, 678.
14 Moss, D.W., Shakespeare, M.J., and Thomas, D.M., *Clin. Chim. Acta*, 1972, **40**, 35.
15 Kadlecová, L., and Štepán, J., *Experientia*, 1972, **28**, 1284.
16 Nealon, D.A., and Henderson, A.R., *J. Clin. Path.*, 1975, **28**, 834.
17 Wroblewski, F., and Gregory, K.F., *Ann. N.Y. Acad. Sci.*, 1961, **94**, 912.
18 McAlpine, P.J., Hopkinson, D.A., and Harris, H., *Ann. Hum. Genet.*, 1970, **34**, 61.
19 Strandjord, P.E., and Clayson, K.J., *J. Lab. Clin. Med.*, 1961, **58**, 962.
20 Dubach, U.C., *Helv. Med. Acta*, 1961, **28**, 469.
21 Wüst, H., Schön, H., and Berg, G., *Klin. Wochenschr.*, 1962, **40**, 1169.
22 Latner, A.L., and Skillen, A.W., *Proc. Assoc. Clin. Biochem.*, 1963, **2**, 100.
23 Nutter, D.D., Trujillo, N.P., and Evans, J.M., *Am. Heart J.*, 1966, **72**, 315.
24 Posen, S., Neale, F.C., and Clubb, J.S., *Ann. Intern. Med.*, 1965, **62**, 1234.
25 Kerkoff, J.F., *Clin. Chim. Acta*, 1968, **22**, 231.
26 Fitzgerald, M.X.M., Fennelly, J.J., and McGeeney, K., *Am. J. Clin. Path.*, 1969, **51**, 194.
27 Tan, I.K., Chio, L.F., and Teow-Suah, L., *Clin. Chim. Acta*, 1972, **41**, 329.
28 Cadeau, B.J., and Malkin, A., *Clin. Chim. Acta*, 1973, **45**, 235.
29 Donaldson, L.A., McIntosh, W.B., and Brodie, M.J., *Scand. J. Gastroenterol.*, 1977, **12**, 637.
30 Donaldson, L.A., Brodie, M.J., McIntosh, W.B., and Joffe, S.N., *Br. J. Surg.*, 1978, **65**, 413.
31 Winkelman, J., Nadler, S., Demetriou, J., and Pileggi, V.J., *Am. J. Clin. Path.*, 1972, **57**, 625.
32 Štepán, J., Volek, V., and Kolář, J., *Clin. Chim. Acta*, 1976, **69**, 1.
33 PetitClerc, C., *Clin. Chem.*, 1976, **22**, 42.
34 Moss, D.W., and Whitby, L.G., *Clin. Chim. Acta*, 1975, **61**, 63.
35 Whitaker, K.B., Whitby, L.G., and Moss, D.W., *Clin. Chim. Acta*, 1977, **80**, 209.
36 Appella, E., and Markert, C.L., *Biochem. Biophys. Res. Commun.*, 1961, **6**, 171.
37 Withycombe, W.A., Plummer, D.T., and Wilkinson, J.H., *Biochem. J.*, 1965, **94**, 384.
38 Birkett, D.J., Conyers, R.A.J., Neale, F.C., Posen, S., and Brudenell-Woods, J., *Arch. Biochem. Biophys.*, 1967, **121**, 470.
39 Butterworth, P.J., and Moss, D.W., *Enzymologia*, 1967, **32**, 269.
40 Hanel, H.K., and Viby-Mogensen, J., *Br. J. Anaesth.*, 1977, **49**, 1251.
41 Brydon, W.G., and Smith, A.F., *Clin. Chim. Acta*, 1973, **43**, 361.
42 Fennelly, J.J., Dunne, J., McGeeney, K., Chong, L., and Fitzgerald, M., *Ann. N.Y. Acad. Sci.*, 1969, **166**, 794.
43 Statland, B.E., Nishi, H.H., and Young, D.S., *Clin. Chem.*, 1972, **18**, 1468.
44 Gerhardt, W., Nielsen, M.L., Nielsen, O.V., Olsen, J.S., and Statland, B.E., *Clin. Chim. Acta*, 1974, **53**, 281.
45 Kim, J.C., Statsny, M., and Manning, J.P., *Clin. Chem.*, 1974, **20**, 816.
46 Conyers, R.A.J., Birkett, D.J., Neale, F.C., Posen, S., and Brudenell-Woods, J., *Biochem. Biophys. Acta*, 1967, **139**, 363.
47 Plocke, D.J., and Vallee, B.L., *Biochemistry*, 1962, **1**, 1039.

48 Applebury, M.L., Johnson, B.P., and Coleman, J.E., *J. Biol. Chem.*, 1970, **245**, 4968.
49 Kutscher, W., and Wörner, A., *Hoppe-Seyler's Z. Physiol. Chem.*, 1936, **236**, 237.
50 Herbert, F.K., *Biochem. J.*, 1944, **38**, xxiii.
51 Abul-Fadl, M.A.M., and King, E.J., *J. Clin. Path.*, 1948, **1**, 80.
52 Peacock, A.C., Reed, R.A., and Highsmith, E.M., *Clin. Chim. Acta*, 1963, **8**, 914.
53 Samuelson, R.C., and Moss, D.W., *Clin. Chim. Acta*, 1976, **72**, 157.
54 Samuelson, R.C., and Moss, D.W., *Clin. Chim. Acta*, 1978, **83**, 167.
55 Scutt, P.B., and Moss, D.W., *Enzymologia*, 1968, **35**, 157.
56 Gazith, J., Schultze, I.T., Gooding, R.H., Womack, F.C., and Colowick, S.P., *Ann. N.Y. Acad. Sci.*, 1968, **151**, 307.
57 Horecker, B.L., in 'Isozymes, I Molecular Structure', ed. Markert, C.L., Academic Press, New York, 1975, p. 11.

4

Immunochemistry of Multiple Forms of Enzymes

Enzymes, like other proteins, stimulate the production of antibodies in animals of species other than those in which they originate. Not infrequently, isoenzymes of a particular enzyme are themselves antigenically distinct, thus providing additional evidence of their different structures. Furthermore, such differences in antigenicity open up a wide range of immunochemical methods of qualitative and quantitative analysis which offer a high degree of specificity and, in some cases, great sensitivity. In many cases, interaction between the enzyme antigen and its specific antibody does not involve the active centre of the enzyme, which remains accessible to the substrate. Therefore, the enzyme–antibody complex is catalytically active. In some instances, the reaction between enzyme and antibody abolishes or greatly reduces catalytic activity, most probably because of steric hindrance. The formation of catalytically active isoenzyme–antibody complexes is an advantage in various methods of immunochemical analysis which depend on the location of the complexes precipitated in gels. Inhibiting antibodies are useful in some forms of analysis in solution, while radio-immunoassay is entirely independent of catalytic activity.

Specificity of Antisera

The applicability of immunochemical methods to isoenzyme analysis depends on the production of specific antisera directed against particular isoenzymes, which in turn requires the availability of suitably purified isoenzyme preparations to serve as antigens. The antigenicity of individual proteins is mostly due to the three-dimensional configuration of relatively small regions of their molecules; variations in structure, *e.g.*, between isoenzymes, which do not affect these regions will not, therefore, introduce differences in antigenicity. However, the immune systems of animals of different species may respond to distinct antigenic determinants when challenged with a particular protein molecule, so that

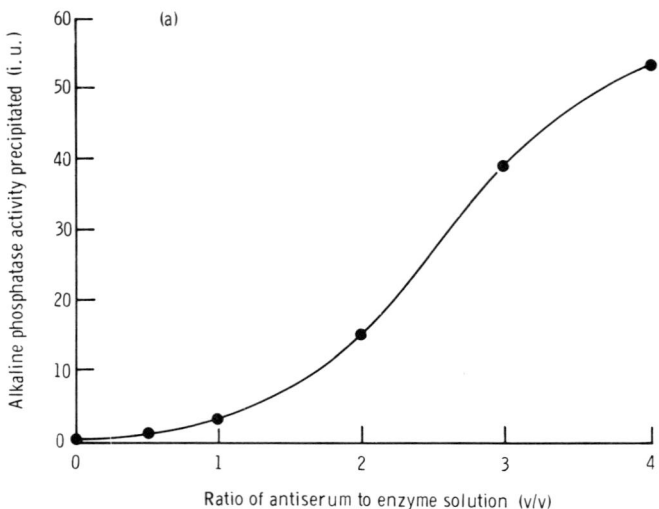

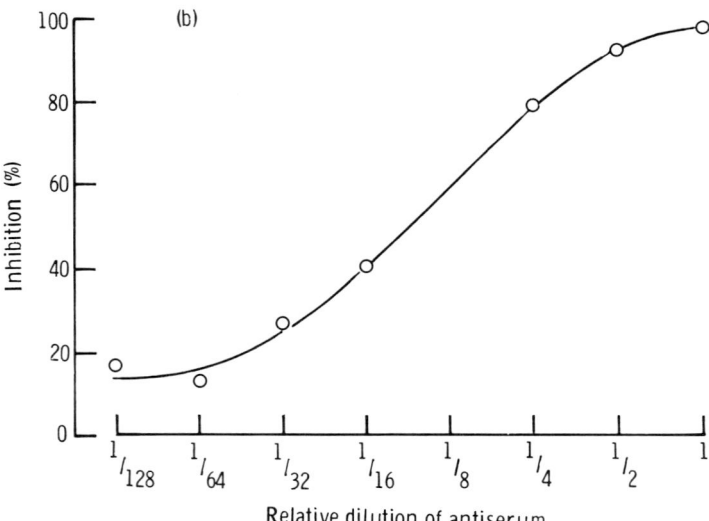

Figure 4.1 (a) *Precipitation of human placental alkaline phosphatase by increasing volumes of a specific rabbit anti-placental phosphatase antiserum*
(b) *Inhibition of the creatine kinase activity of a specimen of human serum, containing predominantly MM-creatine kinase, by an antiserum (E. Merck, Darmstadt) raised against this isoenzyme*

isoenzymes which are antigenically indistinguishable in one species may give rise to different antibodies, and thus to isoenzyme-specific antisera, in animals of a species in which other antigenic determinants are recognised. Similarly, members of a family of isoenzymes from one species may not be antigenically distinct, whereas the corresponding isoenzymes from a different species may stimulate the production of specific antibodies. Examples of inter-species differences in antigenicity are to be found amongst α-amylases from various tissues. Antisera to human pancreatic amylase raised in rabbits did not differentiate between this isoenzyme and human salivary amylase,[1] whereas a rabbit antiserum to porcine pancreatic amylase apparently did not react with amylase from other tissues of the pig.[2]

Antibodies against a particular antigen frequently vary considerably in their specificity and affinity for the antigen from one preparation of antiserum to another. Variations in specificity from one batch of antiserum to another, or heterogeneity of antibodies in this respect, either within a pool of antiserum from several sources or even within a serum from a single animal, can have a marked effect on the extent to which the antiserum cross-reacts with molecules with a close structural relationship to the immunising antigen, *e.g.*, other members of a family of isoenzymes. The specificity of the antiserum for the chosen antigen can be improved by selecting the required antibodies from the rest by a specific separation process such as affinity chromatography, or by adsorbing cross-reacting antibodies with the potential antigens for which they have an affinity; an example of the latter approach is the treatment of antiserum to human intestinal alkaline phosphatase with the antigenically similar placental phosphatase, to increase its specificity for the intestinal isoenzyme.[3] All batches of antiserum should be checked for their cross-reactivity towards potentially interfering antigens, *e.g.*, by immunodiffusion techniques.

Affinity of Antibody for Antigen

The proportion of a fixed amount of antigen (*e.g.*, an isoenzyme) that is bound to antibody typically follows an S-shaped curve when expressed as a function of logarithmically changing antibody concentration (Fig. 4.1). The extent of binding may be determined from the degree of inhibition or precipitation of enzymic activity in the presence of the antibody, or from the incorporation of labelled antigen in competitive-binding assays such as radio-immunoassay. Doubling dilutions of antiserum are frequently chosen as the independent variable, and displacement of the curve towards higher or lower dilutions is a measure of the concentration of antibodies in the antiserum. The maximum dilution of antiserum at which a specified end-point can be reached, *e.g.*, the binding of a given proportion of antigen, is referred to as its titre. Combination of an antigen with its specific antibody is a reversible equilibrium reaction, described by the law of mass action, and the magnitude of the association constant, K, of the antigen–antibody reaction is a measure of the affinity (or avidity) of the antibody for the antigen. Values of K can be estimated from the slope $(= -K)$ of a plot of B/F, the ratio of bound ligand (antigen) to free antigen concentrations, against F, the concentration of free antigen, at constant antibody concentration.

Antisera raised in different animals or at different times generally vary in both affinity and titre. These properties influence the sensitivity of quantitative methods of immunochemical analysis.

Immunochemical Reactions in Solution

In this type of analysis, the isoenzyme antigen is allowed to react with a specific antiserum and the amount of bound isoenzyme is then estimated. The rate of reaction is rapid, depending mainly on the concentrations of the reactants and much less on factors such as ionic strength, provided that the reactants are soluble, or pH within the range of approximately pH 5—8, provided that the ionisation of an essential binding-group is not significantly altered. Increasing temperature accelerates the reaction. However, combination is usually so rapid at room temperature that the increased risk of denaturation of antigen or antibody by incubation at higher temperatures need not be incurred.

When combination with antibody inhibits catalytic activity, the diminution of enzyme activity is a convenient index of the extent of antibody–antigen reaction. The use of inhibiting antisera is valuable in routine analysis of the isoenzyme composition of large numbers of samples, since the process of combination may be essentially complete in a period of minutes and no separation of bound and free isoenzyme (*e.g.*, by centrifugation) is necessary. Enzyme inhibition is not invariably a consequence of combination with antibody, and indeed complete inhibition probably occurs in only a minority of cases. Frequently, however, conditions can be found by varying the relative proportions of antigen and antibody under which the antigen–antibody complex is insoluble and so can be removed, by allowing it to sediment or by centrifugation. The reaction of enzyme and antibody can thus be demonstrated by the partition of catalytic activity between the sediment and the supernatant.

The insolubility of the complex depends on the formation of polymeric structures in which several antibody molecules are linked by divalent molecules of antigen. The relationship between antibody and antigen concentrations is critical in promoting precipitation because, when antigen is present in excess, the binding sites of antibody molecules tend to be occupied by single antigen molecules and few bridges between molecules are formed, while antibody excess favours the formation of complexes consisting at most of only two molecules of antibody linked by antigen. Precipitated complexes tend to redissolve, or even to dissociate, if either antigen or antibody concentration is increased to excess. A titration of antigen with antibody is needed to select the most favourable ratio of these reactants for precipitation to take place. Unlike the initial reaction between antibody and antigen, the formation of insoluble complexes is usually slow, so that a period of incubation at 37 °C of up to one hour, or storage at lower temperatures overnight or for as long as several days, or both, may be necessary for complete precipitation. Complexes which are below the size needed for precipitation may be detectable by nephelometry; however, this technique does not permit the identification of an isoenzyme antigen as a constituent of the complex, which is usually the object of the immunochemical analysis. Precipitation of soluble antigen–antibody complexes can be induced to occur in some cases by the addition of a second antibody, specific for the immunoglobulin with which the primary antigen (*e.g.*, the isoenzyme) has combined. This technique has been applied, for example, to the precipitation of soluble complexes of rat-liver alkaline phosphatase with rabbit antibodies, by means of anti-rabbit γ-globulin antiserum raised in goats.[4] Precipitation with a second antiserum is also useful in searching for complexes which may form between the anti-isoenzyme antiserum and cross-reacting antigens, and which otherwise may escape detection because of the different conditions necessary for precipitation of homologous and heterologous antigen–antibody complexes.

Numerous reports have been made of the use of antigen–antibody reactions in solution as an aid to the differentiation of isoenzymes, and include their application to lactate dehydrogenase,[5,6] creatine kinase,[7] aspartate aminotransferase,[8] alkaline phosphatase,[9,10,11] and other isoenzyme systems.[61,62,65,66] Heteropolymeric isoenzymes usually show immunochemical properties intermediate between those of the respective homopolymers. Placental alkaline phosphatase is antigenically distinct from other human alkaline phosphatases, with the exception of the small-intestinal isoenzyme with which some cross-reaction occurs.[3,10] The common phenotypic isoenzymes of placental alkaline phosphatase possess similar antigenic determinants, and these determinants are also present in the carcinoplacental alkaline phosphatases produced by some tumours. These properties have been exploited in methods for the detection and measurement of placental phosphatase-like isoenzymes in serum which illustrate a variety of types of immunochemical analysis.[12,13,14]

Precipitation with a specific antiserum was used to demonstrate the presence of placental alkaline phosphatase in maternal serum in the last third of normal pregnancy.[12] In a particularly sensitive method, a monospecific anti-placental phosphatase antiserum was polymerised with ethyl chloroformate, rendering the antibody insoluble.[13] Incubation of heat-treated serum samples with the antiserum for one hour was followed by recovery of the now insoluble, but still catalytically active, placental alkaline phosphatase by centrifugation. The enzymic activity of the pellet was determined with a fluorogenic substrate to increase sensitivity still further. Carcinoplacental isoenzyme contributing less than 1% of the total alkaline phosphatase activity of the serum could be determined in this way. A similar technique, in which the phosphatase activity bound to anti-placental phosphatase antibodies which have been polymerised by reaction with glutaraldehyde is determined, was compared with three alternative immunochemical methods: measurement of the inhibition of activity with soluble antibodies, or of residual activity after precipitation with antibodies in either a soluble or a polymerised form.[14] Of these, the immuno-inhibition method was unsatisfactory because of incomplete inhibition of placental alkaline phosphatase activity. The remaining methods gave similar results with elevated isoenzyme activities, but the sensitivity of measurement of activity bound to the polymerised antiserum was superior to that of the alternative methods. High sensitivity has also been shown to be a characteristic of an immuno-assay for human prostatic acid phosphatase, in which the enzyme in serum is bound to a specific antibody coupled to Sepharose 4B. The activity of the bound isoenzyme, recovered by centrifugation, is determined fluorimetrically with α-naphthyl phosphate as substrate.[80] The sensitivity of the method is between one and two hundred times greater than that of equivalent radio-immunoassays.

Antisera which can distinguish consistently between the alkaline phosphatases derived from bone or liver have not been readily available, so that immunochemical methods have not so far contributed significantly to what is clinically the most important problem in the analysis of mixtures of alkaline phosphatase isoenzymes in serum. Although immunochemical methods for determining isoenzymes of other enzymes such as aldolase and lactate dehydrogenase have been described,[15, 16] the most widespread diagnostic application of antisera against specific isoenzymes in the quantitative analysis of human serum samples by immunochemical reactions in solution is currently to be found in the determination of the isoenzymes of creatine kinase.

Antisera raised in rabbits to either purified MM- or BB-creatine kinase isoenzymes were each found to be specific for their respective homologous antigens, with no cross-reaction with the heterologous isoenzyme. Both antisera interacted with the hybrid MB-creatine kinase.[17] The extent of combination of creatine kinase isoenzymes in tissue extracts or serum samples with the antisera was estimated from the residual enzymic activity, either after incubation for successive periods of 30 min at room temperature and 4 °C (immuno-inhibition), or after centrifugation following these incubations (immunoprecipitation). Activity of the MM isoenzyme was completely inhibited by its specific antiserum. However, BB-creatine kinase activity was inhibited to the extent of about 88% by anti-BB antiserum, although the residual activity was completely removed from solution by centrifugation. The MB isoenzyme was inhibited to the extent of about 80% by either antiserum, with 10% of the original activity remaining in the supernatant after centrifugation; *i.e.*, the hybrid isoenzyme reacted incompletely with a single antiserum. Addition of the second antiserum was needed to precipitate the MB isoenzyme completely. The isoenzyme composition of samples containing creatine kinase was calculated from the percentage residual activities, a and b, measured after precipitation with anti-MM and anti-BB antisera in the following way:

$$a = \% \text{ BB} + 10\% \text{ MB}$$

$$b = \% \text{ MM} + 10\% \text{ MB}$$

$$\text{and } 100 = \% \text{ BB} + \% \text{ MM} + \% \text{ MB}$$

$$\text{therefore, } \% \text{ MB} = 1.25(100 - a - b)$$

The results of this type of analysis agreed well with the known compositions of mixtures of the three isoenzymes.

This and similar[18] immunoprecipitation procedures are time-consuming compared with methods based on inhibition alone, and are effective over a more limited range of enzyme activity. If the object of analysis is to determine the content of MB-creatine kinase, as is most frequently the case in diagnostic laboratories, a further simplification can be achieved by the use of a single antiserum, if it is assumed that B subunits are present only as components of the MB isoenzyme. Two methods have been described, both of which employ a human MM-creatine kinase-inhibiting antiserum prepared in goats.[19,20] Inhibition of MM-creatine kinase was shown to be essentially complete within 3 min of adding this antiserum, with 50% inhibition of the MB dimer and no effect on the BB isoenzyme.[19] The value of $t_{\frac{1}{2}}$ for the reaction with antibodies was estimated to be 38 s at 37 °C in a later study and the final degree of inhibition of MM-creatine kinase 99%.[20] There is some inhibition of residual activity of the B subunit of MB-creatine kinase as a result of combination of the M protomers with anti-MM antibodies, amounting to about 5%. Therefore, doubling the measured residual activity after immuno-inhibition of a mixture of MM- and MB-creatine kinases underestimates the original activity of the latter isoenzyme by up to 10%. Methods of this type are designed to detect and measure an isoenzyme which is a minor component of the sample; methods of assay of enzyme activity must be selected, therefore, which are capable of reliably determining low residual activities. A sensitive ATP-dependent luciferase reaction has been used in one immuno-inhibition method for creatine kinase isoenzymes.[21] For this enzyme, also, an effective activator (*e.g.*, *N*-acetyl cysteine) is required. Furthermore, blank reactions which can be neglected when measuring total enzyme activity may be comparable with the remaining, uninhibited isoenzyme activity, so that correction of the residual reaction rate is needed. For example, suppression of the interfering activity of adenylate kinase by inhibitors is adequate when measuring total creatine kinase activity, but not for all samples when immuno-inhibition methods are used to determine the MB isoenzyme.[20] Coefficients of variation of replicate analyses by these methods are of the order of ± 5—10%, depending on the level of activity of B subunits, and MB-creatine kinase activities as low as 3—4 i.u. l^{-1} at 25 °C can be measured.

Reactions between isoenzymes and specific antisera have been used as a preliminary to zone electrophoresis, to modify the migration of the isoenzymes or to aid in the identification of particular zones. Examples are the electrophoresis of supernatant solutions after the precipitation of isoenzymes of alkaline phosphatase with antisera,[10, 22] and the use of anti-BB-creatine kinase antiserum to distinguish between the BB isoenzyme and a fluorescent artefact of similar electrophoretic mobility.[23]

Precipitation Reactions in Gels

Gels provide a favourable environment for the precipitation of antigen–antibody complexes, by reducing convection currents and thermal agitation which may disrupt the growth of the lattice of antigen and antibody molecules. Diffusion of the antigen and antibody towards each other in gels generates concentration gradients of each which, at some point in their intersection, usually provide the relative concentrations appropriate for precipitation. When this process has begun, migration towards the precipitation zone of further molecules of antigen and antibody increases the amount of precipitated complex. However, the passive diffusion of large molecules through gels is slow, so that the initial reaction between antigen and antibody is retarded and a long period is usually needed to accumulate significant amounts of precipitate. Low temperatures (*e.g.*, 4 °C) are often chosen for diffusion in gels; although rates of diffusion are reduced compared with those at higher temperatures, the insoluble complex is more stable. Diffusion of antigen and antibody through the gel is accelerated in several methods by applying an electric potential.

Inactive enzyme–antibody precipitates can be located after diffusion in gels by staining with dyes which are taken up by proteins, or, if the concentration of the antigen is high enough, by the appearance of opalescent precipitin lines in the clearer gel matrix. However, these methods do not postively identify specific isoenzyme precipitates. This can be achieved, together with increased sensitivity, in the case of enzyme–antibody reactions in which catalytic activity is retained, since individual precipitates can be identified after immunodiffusion by specific staining reactions similar to those used to locate isoenzyme zones after electrophoresis.[24]

Immunodiffusion.—In the various applications of this technique, the antibody and antigen move through the gel by passive diffusion. Agarose is now almost invariably chosen to prepare the gel, because of the clarity, purity, and mechanical strength of its gels at low concentrations, and the absence of negatively charged groups which may retard the migration of basic proteins. Concentrations of approximately 0.8% m/V of agarose are usually used, giving pore sizes which do not form a barrier to larger molecules. Gels can be prepared in saline or in dilute buffer solutions. The choice of a buffer will depend to some extent on the nature of the enzyme reaction subsequently to be carried out in the gel; if the nature of the buffer is such that it may interfere with formation of the gel, or with combination of antibody and antigen, it can be introduced at a later stage in the procedure by soaking the gel in it. Agarose solutions are prepared by heating the polysaccharide to about 100 °C in the solvent and become gels at a temperature of 40—45 °C. The gelation temperature varies with different agarose preparations. Those gelling at lower temperatures have the advantage that antibodies incorporated just before the gel forms are less likely to be denatured.

Although now largely supplanted by agarose, agar gels have been extensively used in immunodiffusion. The electro-osmosis which occurs in these gels can be turned to advantage in some electro-immunodiffusion techniques. Various other gels which have been employed as supporting media in zone electrophoresis, such as starch or polyacrylamide gels, have also been applied to immunodiffusion methods, as has cellulose acetate, but without acquiring the popularity of agarose or agar. Immunodiffusion is now almost invariably carried out in a layer of gel 1—2 mm thick on glass plates, older techniques in which the gel is formed in a small test-tube being seldom used. In single diffusion tests, one of the reactants, usually the antigen, diffuses into a gel which typically contains the antibody. Double-diffusion methods involve the migration of both components of the reaction towards each other in the gel.

Radial diffusion of antigen from wells 1—2 mm in diameter into an agar or agarose gel layer containing antiserum to form precipitin rings (the Mancini technique) is a simple method of estimating antigen concentration. The diameter of the precipitin rings is logarithmically related to the concentration of antigen, so that a calibration curve can be constructed if an antigen solution of known concentration is available. However, the method is slow, requiring 48—72 h for development of the rings and is often imprecise because of the difficulty of measuring their diameters, especially if they are faint or irregular, and it appears to have been used only rarely for determination of isoenzymes.[15,50]

In double immunodiffusion tests, antigen and antibody solutions are introduced into separate wells in an agar- or agarose-gel-coated plate and diffusion of both produces a precipitin line. After washing to remove unprecipitated material, the precipitated antigen–antibody complex is stained, either for protein or, preferably, for enzymic activity when active

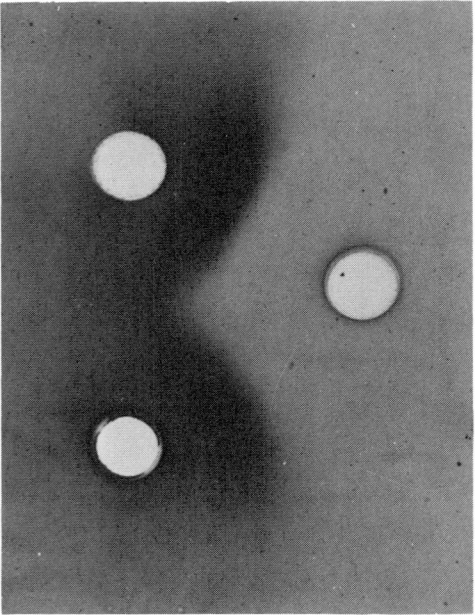

Figure 4.2 *Ouchterlony double immunodiffusion experiment showing the antigenic identity of a carcinoplacental (Regan) isoenzyme of alkaline phosphatase in a human serum sample (in well at lower left) with the normal placental isoenzyme (upper left). The well at the right contained anti-placental phosphatase antiserum. The still catalytically active isoenzyme-antibody precipitates were stained as in Fig. 2.3*

complexes are present. Several antigen wells are arranged in a circle with a single antiserum well at the centre in the Ouchterlony technique; this allows the reactivity of several potential antigens towards a single antiserum to be compared, undisturbed crossing of precipitin lines between the antiserum and different antigens indicating non-identity, incomplete fusion with spur formation partial cross-reaction, and complete fusion of the lines antigenic identity (Fig. 4.2). The Ouchterlony double-immunodiffusion test or variants of it have been used extensively in characterising antisera raised against specific isoenzymes and to investigate the antigenic relationships between isoenzymes.[3, 17, 24−28, 60, 61, 62, 66, 68, 71]

An important variant of double immunodiffusion is *immunoelectrophoresis*. In this technique, longitudinal separation of a mixture of antigens by electrophoresis is followed by lateral diffusion of the separated zones, to meet and react with antibody diffusing from a trough cut parallel to the direction of electrophoretic migration. In contrast to the electro-immunodiffusion methods described below, the application of the electric field is a preliminary to the conjunction of antigen and antibody, which takes place by passive diffusion. The advantage of immuno-electrophoresis which has resulted in its extensive use in isoenzyme analysis is that the antigenic similarities or dissimilarities of electrophoretically characterised enzyme zones can be determined simultaneously with a single sample of antiserum.[24, 25, 26, 28, 29, 62, 66, 68, 71] Moreover, since diffusion from the separated zones takes place in both lateral directions, two different antisera can be used after each separation, one on each side of the zones, further facilitating immuno-chemical comparisons.

A single gel, typically agar or agarose, is frequently used for both the initial electrophoretic separation and the subsequent immunodiffusion.

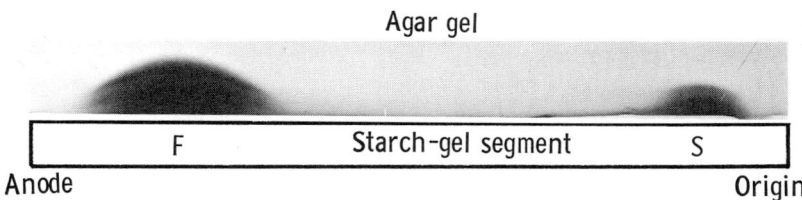

Agar gel

F | Starch-gel segment | S

Anode | Origin

Figure 4.3 *Immuno-electrophoresis of alkaline phosphatase components in an extract of human placenta. Electrophoresis in starch gel at pH 8.6 to separate the differently sized F and S components was followed by diffusion into agar gel containing an antiserum raised against zone F, with which the slow component also reacts. The precipitin arcs were stained as in Fig. 2.3.* [Reproduced, with permission, from: D. W. Moss, *Clin. Chim. Acta* (Elsevier), 1973, **43**, 447]

However, in addition to this and other single-gel systems, *e.g.*, using large-pore polyacrylamide gel, combinations of different gels for the stages of electrophoresis and immunodiffusion may be useful, particularly when more concentrated gels of starch[30] or polyacrylamide[31] are needed to separate enzyme zones of different molecular sizes by molecular sieving during electrophoresis. Passive diffusion of large molecules in these small-pore gels is so slow that they are unsuitable as media for immunodiffusion processes. Therefore, a longitudinal slice of the electrophoresis gel, or the whole gel in the case of electrophoresis in small cylinders of polyacrylamide gel, is placed on a glass plate on which molten agar or agarose gel is poured to form a gel layer surrounding and in contact with the electrophoresis gel. After the agarose gel has solidified, a trough is cut parallel to the electrophoresis strip to receive the antiserum (Fig. 4.3).

The practical considerations in the first stage of immunoelectrophoresis are similar to those applying generally to zone electrophoresis in gels (Chapter 2). In the subsequent immunodiffusion phase, important factors are the spacing between the centre line of the separated enzyme zones and the antiserum trough, and the dilutions of the antiserum and antigen solutions to be used. If these factors are not optimal, precipitation may take place too close to the antiserum trough to allow significant irregularities, crossing or fusion of precipitin arcs to be discerned. Similarly, leakage of antisera from the trough can blur precipitin arcs, and irregularities or splits in the walls of the troughs may result in distortions of precipitin arcs which may simulate various types of immunochemical interactions. Given carefully controlled and appropriate conditions, however, the shapes of precipitin arcs developed in immunoelectrophoresis can provide much information about the antigenic relationships between enzyme zones.

Electrically assisted Immunodiffusion.—As has already been remarked, immunoprecipitation resulting from passive diffusion in gels is a slow process, hours or even days being required to achieve detectable precipitation. Methods in which diffusion is accelerated by applying an electric field not only have advantages of rapidity but also are more suitable for quantitative analysis than analogous passive-diffusion procedures. Both single and double electro-immunodiffusion techniques have been applied to the study of multiple forms of enzymes. In the former, as in single immunodiffusion without the aid of an applied electric field, the antigen under study migrates through a gel containing the stationary complementary antibody, or, much more rarely, antibody molecules travel through a field of antigen. In double electro-immunodiffusion (often called counter immuno-electrophoresis), both antigen and antibody molecules migrate under the influence of the electric field.

The condition that the antibody molecules should remain essentially stationary during the migration of antigen in electrically assisted single immunodiffusion imposes limitations on the pH at which the process is carried out. Although a small movement of antibody molecules under the influence of the electric field may not affect the results, a significant degree of electrophoretic migration will result in an uneven distribution of antibody molecules in the gel so that quantitative assessment is impaired. However, antibody molecules generally have little net charge at alkaline pH values, such as pH 8.6, in contrast to many protein antigens, including isoenzymes, which exhibit appreciable electrophoretic mobility under these conditions. The effects of small inequalities in antibody distribution resulting from movement of antibodies during electrophoresis can be overcome in quantitative analysis by the inclusion of standard solutions of antigen amongst the unknown samples. When the antigen and antibody have similar ionisation characteristics, it becomes difficult or impossible to select pH conditions under which their electrophoretic mobilities are sufficiently different for electroimmunodiffusion to be practicable. In such cases, useful differences in net charge at a given pH can sometimes be created by chemical modification of one or other component of the system (usually the antigen). Needless to say, such modification must not materially alter the antigenic identity of the molecule, nor should it destroy the catalytic activity of an enzyme antigen, if this property is to be used to locate the antigen–antibody precipitate. Acetylation or carbamoylation of the free amino groups of proteins prevents their ionisation to positively-charged NH_3^+ residues, thus increasing the net negative charge of protein molecules and their electrophoretic mobility towards the anode. Carbamoylation with cyanate has been used to improve the migration of isoenzyme C (isoenzyme II) of human carbonic anhydrase (EC 4.2.1.1) during electro-immunoassay.[32]

Rocket immuno-electrophoresis, also known as Laurell electro-immunoassay, can be regarded as a variant of the Mancini radial immunodiffusion technique in which antigen molecules are drawn from their wells through the antibody-containing gel under the influence of an electric field.[33] As in passive diffusion, migration continues until antigen–antibody complexes are precipitated. The distance into the gel at which precipitation occurs depends on the antigen concentration, since the concentration of antibody is initially constant throughout the gel; thus, the height of the rocket-shaped precipitin peak is a measure of antigen concentration. The peak heights given by antigen solutions of unknown concentration are compared with those of standard antigen solutions run simultaneously. Relative concentrations of different antigen solutions are more accurately reflected by the areas enclosed by their respective rockets, rather than simply by their heights, but any advantage thus obtained is more than offset by the practical difficulties of measurement of the areas. Even with simple methods, peak heights of the order of 1—5 cm can be measured with an accuracy of about 0.2 mm and, although the overall precision of the method depends on various additional factors, coefficients of variation of less than ±7% have been found for replicate determinations of placental alkaline phosphatase by rocket electro-immunoassay.[34] The sensitivity of the method, also, is the resultant of several factors. The distance of migration should be great enough for the peak height to be accurately measurable, and can be increased for dilute antigen solutions by reducing the concentration of antiserum in the gel. Enough antigen–antibody complex must be present at the tip of the rocket to be located by staining or other means, and low concentrations of the two components means that the concentration of their complex at the point of precipitation will in turn be low. Although the detection of weak rockets by conventional protein stains may be impossible, catalytically active enzyme–antibody precipitates can be detected with great sensitivity, as in other forms of immunodiffusion analysis.

The antiserum used in rocket immunoassay ideally should not cross-react with isoenzymes related to that under study, since identification of a particular isoenzyme by its electrophoretic mobility is not possible in this one-dimensional technique, as it is in classical immuno-electrophoresis or 'crossed' immuno-electrophoresis. Some degree of cross-reaction which involves a minor isoenzyme component and which therefore produces small, faint rockets may not interfere with the identification and measurement of the selected isoenzyme. However, if the proportions of heterologous and homologous antigens are comparable, the ability to discriminate between them may be lost.

Apart from the uniform distribution of a suitable antiserum throughout the gel, other important practical considerations in rocket electro-immunoassay include the application of a uniform electrical field, to prevent distortion of the peaks; this requires good electrical connections to the buffer reservoirs across the whole width of the anode and cathode ends of the gel, and a gel layer of even thickness (typically about 1 mm). The gel (usually 0.8—1% m/V agarose) is made in a suitable buffer solution, *e.g.*, sodium barbitone (0.2—1 mol l^{-1})– barbituric acid, pH 8.6. The more dilute buffer solutions reduce heat production during electrophoresis and aid precipitation, but have lower buffering capacity. Voltage gradients of 2—8 volts cm^{-1} are usually employed, development of the rockets taking from more than 8 h to less than 2 h, according to voltage. Unprecipitated proteins are removed by washing with saline solution or water before staining.

Rocket electro-immunoassay has been applied to the analysis of several enzymes and isoenzymes, including carbonic anhydrase,[32] placental alkaline phosphatase[34] (Fig. 4.4), prostatic acid phosphatase,[35]

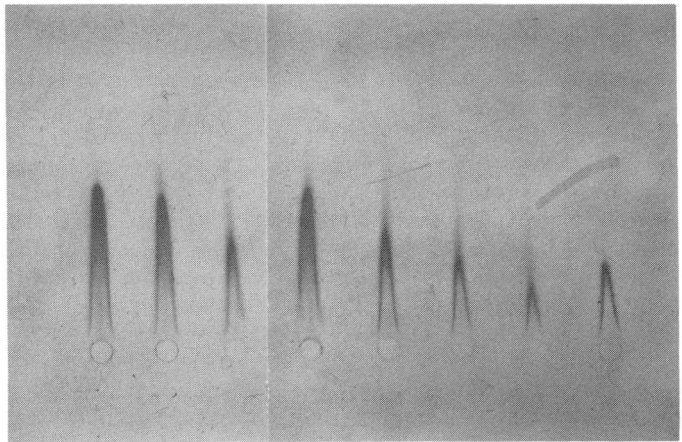

Figure 4.4 *'Rocket' electro-immunoassay of human placental alkaline phosphatase in serum. The range of activity of the samples was from approximately 0.8 to 12 times the upper limit of normal for total alkaline phosphatase activity in the serum of adults. The peaks were located as in Fig. 2.6.* [Reproduced, with permission, from: D. T. Forman, D. W. Moss, and K. B. Whitaker, *Clin. Chim. Acta* (Elsevier), 1976, **68**, 287]

glucose-6-phosphate dehydrogenase in human leukaemias,[36] and several hydrolytic enzymes in a human diploid cell line.[37]

Crossed immuno-electrophoresis combines the separation and iden- tification of isoenzyme zones given by electrophoresis with the rapid and quantitative immunochemical analysis characteristic of rocket electro- immunoassay.[38] A preliminary electrophoretic separation of the mixture of antigens (*i.e.*, the enzymes or isoenzymes) is followed by a second migration of the zones in a direction at right angles to their original movement into an antibody-containing gel, again under the influence of an electric field. Agarose gel (*ca.* 1% *m/V*) is usually chosen as the medium for both migrations, with conditions of buffer composition and pH, voltages, and antibody concentration in the second gel, selected according to considerations similar to those obtaining in one-dimensional immuno-electrophoresis and rocket electro-immunoassay. In earlier descriptions of crossed immuno-electrophoresis, the original sample was applied into a slit in the gel cut transversely to the direction of electrophoretic migration, and a narrow strip was cut longitudinally from the centre of the strip after electrophoresis to serve as the sample gel from which the second migration originates. However, since such a strip contains variable proportions of the several antigens, this procedure is less suitable for quantitative analysis than later versions in which electrophoresis originates from a small, circular well containing the sample, the full width of the electrophoresis zones being included in the second migration. In either case the gel strip is transferred to a second glass plate on to which the molten antibody-containing gel is poured so as to make contact with one of the long edges of the electrophoresis gel.

After migration is completed, unprecipitated proteins are removed by washing and the precipitin peaks are located, preferably by specific reactions dependent on enzymic activity. The relative amounts of different antigens are proportional to the areas under their respective peaks. Absolute amounts of antigens can be determined if solutions of known concentration are available for calibration purposes. Areas are measured by planimetry or, less accurately but more simply, by multiplying the height of the peak by its width at half the peak height. Although its quantitative aspect is probably the greatest single advantage of crossed immuno-electrophoresis over conventional immuno-electrophoresis, interactions of various kinds between the precipitin lines produced by different antigens are also more easily studied in the well-developed peaks of the crossed technique than in the arcs of immuno-electrophoresis.

Crossed immuno-electrophoresis has been used to determine the amounts of two α-amylase isoenzymes during germination of barley seeds.[39] In a study of the reactions of a variety of enzymes in microsomes and plasma membranes from rat liver with antisera to each of these cell fractions, the greater resolving power of crossed immuno-electrophoresis was demonstrated by the recognition of ten antigens with uridine-diphosphatase activity in the microsomal fraction by this technique, compared with only three which could be resolved by conventional immuno-electrophoresis.[40, 41] Specific enzymic staining reactions were used to identify the antigen–antibody precipitates in all these investigations.

Counter immuno-electrophoresis is an electrically assisted variant of double immunodiffusion in which both antigen and antibody migrate in the electric field. To ensure that they meet, conditions must be chosen under which the antigen and antibody move in opposite directions. Ideally, therefore, the pH of the medium should be between the respective isoelectric pH values of these two components, so that one moves towards the cathode and the other towards the anode. However, the isoelectric points of antibodies are in the region of pH 5—6 and, although somewhat higher, are not markedly different from the corresponding values for many enzymic and isoenzymic antigens, so that this condition is rarely achieved in practice. Nevertheless, the principle of counter immuno-electrophoresis can be applied under circumstances in which the antigen migrates electrophoretically towards the anode while the antibody moves in the reverse direction largely by electro-endosmosis.

Although counter immuno-electrophoresis is frequently carried out in agarose gel (0.75—1% m/V in buffer), the merits of agarose compared with agar are less apparent in this technique than in other types of immunodiffusion; indeed, agar has a distinct advantage in most applications because of the increased electro-endosmotic flow which it promotes.[42,43] The antiserum and antigen solution are introduced into wells (*ca.* 3 mm dia.) cut in the gel layer and separated by a distance of about 1 cm in the direction of migration, with the antiserum in the more-anodal well and the antigen solution in the well nearer to the cathode. After electrophoresis for 1—2 h, depending on voltage, pH, and buffer concentration, the antigen–antibody complex precipitated between the wells is made visible by staining for protein or enzyme activity. The time of electrophoresis is critical since over-running may cause the precipitate to re-dissolve in excess of antibody or antigen. The method is generally used for rapid quantitation of an antigen and this can be achieved by determining the greatest dilution of the antigen solution which produces a detectable precipitate.

Two counter immuno-electrophoretic assays have been described for the prostatic isoenzyme of acid phosphatase in human serum.[44,45] Both employ agarose gel as the diffusion medium, with naphthol AS-MX phosphate or α-naphthyl phosphate as substrates for the determination of the enzymically active immunoprecipitate, the liberated naphthols being coupled with stabilised diazonium salts (Fast Red Violet LB or Fast Red Garnet GBC) to form coloured precipitates. However, in one method,[44] a barbitone buffer (0.05 mol l^{-1}) at pH 8.6 was chosen, whereas, in the other,[45] phosphate buffer (0.05 mol l^{-1}) at pH 6.5 was preferred. The isoelectric point of purified prostatic acid phosphatase has been estimated to be close to pH 4.9 by isoelectric focusing;[46] therefore, the isoenzyme still has a significant net negative charge at pH 6.5, and the use of this pH rather than pH 8.6 would seem to favour the contrary movement of antigen and antibody by reducing the net negative charge on the latter while maintaining a significant charge on the former. However, precipitin lines between the two wells are obtained in both methods. Both are also of comparable sensitivity (minimum detection limit, 20—30 μg l^{-1}) for the detection of the purified isoenzyme, but, in the method at higher pH, sensitivity for the enzyme in serum was lower than this by a factor of about tenfold, possibly because of interference with the staining reaction by serum proteins migrating with acid phosphatase.[44] As studies with other antigens have also indicated, these counter immuno-electrophoresis methods are significantly less sensitive than corresponding radio-immunoassays.

Many ingenious variations and extensions of electrically assisted immunodiffusion techniques have been described, designed generally to improve the recognition of specific antigens and to facilitate the study of cross-reactions between antigens. For example, close spacing of sample wells in rocket electro-immunoassay or crossed immuno-electrophoresis allows fusion of the precipitates formed by neighbouring, cross-reacting antigens to take place. A similar effect can be obtained in counter immuno-electrophoresis by the use of two antigen wells side by side, while a preliminary electrophoretic separation of the antigen solution followed by counter immuno-electrophoresis of antiserum from a trough parallel to the original direction of migration converts the counter immuno-electrophoretic technique into an electrically accelerated version of conventional immuno-electrophoresis.

Radio-immunoassay

This particularly sensitive type of analysis has come to occupy an important place in the determination of a wide range of antigens, haptens, and antibodies, and a significant number of radio-immuno-assays of isoenzymes have now been described. Radio-immunoassays are based on competition for specific binding sites of antibody molecules between the antigen being determined and a known amount of the same antigen labelled with a radioactive isotope. Thus, for the application of radio-immunoassay to isoenzyme analysis, a quantity of the purified isoenzyme is required, not only for the production of a specific antiserum, but also for radioactive labelling and to serve as a calibration standard in the assay. The specificity of the assay is, of course, determined by the specificity of the antibodies. Sensitivity is dependent on the titre of the antiserum and the affinity of its antibodies for the antigen: given an antiserum which is satisfactory in these respects, radio-immunoassays are able to determine amounts of antigen in the nanogram range or below, because of the ability to extend the counting time to register a significant number of disintegrations.

The ability to distinguish between bound and free antigen is an essential requirement of competitive-binding assays. The two phases must be separated physically in radio-immunoassay, although this is not necessary in certain other competitive-binding methods. Antigen–antibody complexes do not precipitate at the low concentrations typical of radio-immunoassay, but can be induced to do so by addition of a second antiserum, directed against the immunoglobulin comprising the first antibody. Alternatively, the specific antibodies are attached to a solid matrix such as a cross-linked dextran or cellulose, either covalently, or simply by non-specific adsorption, *e.g.*, to the walls of a plastic test-tube. The antibodies themselves may be chemically polymerised to render them insoluble. Various separation methods used in other applications of radio-immunoassay, such as filtration, or precipitation with salts or organic solvents, are less able satisfactorily to separate bound and unbound enzymes because of their general similarities in properties, resulting, for example, in non-specific precipitation.

Radioactive iodine, in the form of the isotope ^{125}I (half-life 60 days), is almost invariably used to label enzyme and isoenzyme antigens for radio-immunoassay. Radioactive iodide is oxidised to iodine by acid iodate solution, lactoperoxidase and hydrogen peroxide, or chloramine-T, and the iodine atoms are incorporated into tyrosine side-chains of the enzyme molecule. Unreacted iodine is destroyed by reduction with sodium metabisulphite. Exposure of enzyme molecules to possibly damaging oxidising conditions can be avoided by first iodinating N-succinimidyl-3-(p-hydroxyphenyl) propionate, which then forms covalent linkages with the amino groups of the enzyme antigen. The labelled antigen must not only retain those features of its structure which are recognised by the antibodies, but also should have an affinity for the antibodies similar to that of the unlabelled molecule, or the sensitivity of the assay may be reduced. A study of radioactively labelled MM and BB isoenzymes of creatine kinase showed that the extent of immunoprecipitation of the isoenzymes by their complementary antisera was not affected by a fourfold increase in the degree of labelling,[47] demonstrating that, in this case at least, the process does not markedly affect the antigen–antibody reaction. In addition, the labelled preparation must be free from unlabelled antigen and from labelled or unlabelled degradation products. After labelling, therefore, the purity of the antigen is usually checked and increased if necessary by electrophoretic or various chromatographic procedures. These processes also remove unreacted labelling reagents. Labelling may or may not affect the catalytic activity of an enzyme antigen, depending on the relationship between the labelled side-chains and the active centre. However, although persistence of catalytic activity may provide evidence that labelling has not damaged the enzyme molecule, the sensitivity of radio-immunoassay does not depend on catalytic activity, in contrast to immuno-diffusion assays, the sensitivity of which is increased by the ability to apply specific stains for enzymically active immunoprecipitates.

Radio-immunoassays have been described for carbonic anhydrase isoenzymes,[48] type II carboxypeptidase B (EC 3.4.2.2) from human pancreas,[49] porcine pancreatic amylase,[2] human hexosaminidases A and B,[50] the prostatic isoenzyme of human acid phosphatase,[26,51] human placental and carcinoplacental (Regan) alkaline phosphatases,[31,52,53,54] and the M and B subunits, and thus the three isoenzymic dimers, of creatine kinase from human tissues.[55-59] All these methods make use of ^{125}I as the radioactive label and in most cases the double-antibody technique, in which the antibody-bound isoenzyme is precipitated with an anti-γ-globulin antiserum and recovered by centrifugation, is employed to separate the bound and unbound phases. Anti-isoenzyme antibodies polymerised with ethyl chloroformate and homogenised to fine particles form an insoluble binding phase in one assay of carcinoplacental alkaline phosphatase, centrifugal sedimentation of the bound fraction of the antigen being further aided by the addition of an homogenised starch-gel suspension.[53] The bound and free antigen fractions have been separated by the solid-phase technique in some assays, by coating polypropylene tubes with the anti-isoenzyme antiserum.[26,31] Treatment of the tubes with bovine serum albumin after they have been coated with antiserum saturates any remaining non-specific protein-binding sites. Placental alkaline phosphatase has been assayed by a combined double-antibody and solid-phase technique, in which the anti-γ-globulin antibodies are added in the form of insoluble cellulose particles coated with these antibodies to aid subsequent centrifugal separation of the antibody-bound phase.[54]

These methods display the great sensitivity characteristic of radio-immunoassay. However, incubation periods of several hours are often required, either for the equilibration of the isoenzyme present in the sample and its labelled counterpart with the specific antiserum, or for the subsequent separation reactions, or both. Although this does not represent a serious drawback in assays carried out for non-clinical purposes, or in clinical assays for isoenzymes such as prostatic acid phosphatase or placental alkaline phosphatase, it can represent a marked disadvantage of radio-immunoassay compared with other methods of determining amounts of the isoenzymes of creatine kinase in human serum, when the clinical need is often for a rapid confirmation of the occurrence of myocardial infarction.

Several alternatives to labelling with radioactive isotopes in competitive-binding assays are now receiving increasing attention, in order to avoid disadvantages associated with the use of radioactive materials. The labelling of antigens by coupling enzyme molecules to them is particularly attractive, because of the sensitivity with which the catalytic activity of the label can be detected and also because, in those cases in which binding to the antibody is accompanied by inhibition of the enzyme label, physical separation of the bound and free phases is not required. This approach has not yet been applied to instances in which the antigen being determined is itself an enzyme, *e.g.*, a particular isoenzyme.

Other Aspects of Immunochemical Studies of Isoenzymes

Immunochemical methods provide means of exploring aspects of isoenzyme structure which may be inaccessible to approaches such as separations based on net molecular charge, or comparisons of catalytic activity. The value of immunochemical methods is particularly apparent in the investigation of inherited enzyme variants, or of altered activities of enzymes arising in the course of disease or in response to inducing agents. In these cases, quantitative changes in the level of a particular enzyme activity may occur without significant alterations in various parameters, such as Michaelis constant or pH dependence, which are characteristic of the catalysed reaction. Indeed, the level of activity may be so low that these properties cannot be measured, nor can the resistance of activity to various denaturants, while other characterisation procedures become equally impracticable. Therefore, the possibility offered by immunochemical techniques of measuring the amount of enzyme protein (or enzyme-like protein) present in the modified system is correspondingly important. Immunochemical assays have been used to demonstrate the production by rare allelic genes of isoenzymes with reduced or increased specific activity compared with the products of the corresponding usual genes. Some alleles responsible for certain enzyme-deficiency states and inborn errors of metabolism give rise to products (termed cross-reacting materials) which, while totally devoid of enzymic activity, nevertheless retain antigenic determinants which are recognised by antisera raised against the normal isoenzyme.[60-71]

Two apparently different mechanisms resulting in an increased alkaline phosphatase activity have been distinguished with the aid of immunochemical titrations. Ligation of the common bile duct of the rat is followed by a rapid rise in alkaline phosphatase activity in liver tissue, which was shown immunochemically to be accompanied by an increase in the amount of enzyme protein.[4] In contrast, however, an increased amount of enzyme protein could not be demonstrated during the enhancement by cortisone of the alkaline phosphatase activity of HeLa cells.[72]

Although isoenzyme molecules are themselves potent antigens, association of isoenzymes with non-protein moieties, or with other proteins in the form of complexes, may introduce new sets of antigenic determinants which are themselves capable of interacting with specific antibodies. These possibilities may complicate the interpretation of the apparent immunological properties of isoenzymes but, on the other hand, they may provide new opportunities to investigate the nature of binding phenomena involving isoenzymes. Several enzymes have been shown to appear occasionally in blood plasma in the form of high molecular weight complexes, and immunochemical methods have played an important part in determining the nature of these distinctive enzyme forms. Association of amylase in serum with immunoglobulins gives rise to the condition of macro-amylasaemia. The protein associated with the enzyme is either immunoglobulin A or G,[73, 74, 75] and the identity of the immunoglobulin and its presence in the complex can be demonstrated by immuno-precipitation, immunodiffusion-analysis, or immuno-

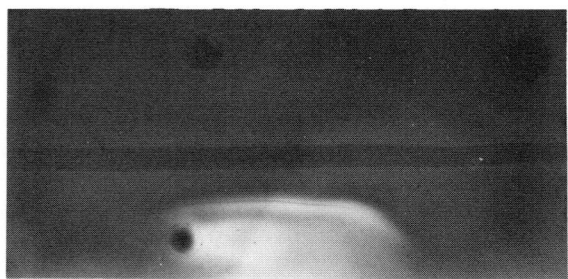

Figure 4.5 *Cathodal portion of an immuno-electrophoretic separation in agar gel of macroamylase in human serum. The enzyme, complexed with the immunoglobulin* IgA *in this case, has been precipitated by an anti-human* IgA *antiserum diffusing from the trough. Amylase activity associated with* IgA *is shown by the white area surrounding the precipitin arc, after overlaying the electrophoresis gel with a second agar gel containing soluble starch and subsequently immersing it in an iodine solution. The precipitated immunoglobulin itself is visible as a thin, dark line in the centre of the arc. The upper well contained a serum with elevated free amylase activity.* (By courtesy of Dr. H. G. M. Freeman)

electrophoresis, with the aid of appropriate antisera (Fig. 4.5). Both these immunoglobulins have also been identified immunochemically in complexes with lactate dehydrogenase,[76, 77] and association of alkaline phosphatase with immunoglobulin G has been reported.[78, 79]

References for Chapter 4

1 Carney, J.A., *Clin. Chim. Acta*, 1976, **67**, 153.
2 Ryan, J.P., Appert, H.E., Carballo, J., and Davies, R.H., *Proc. Soc. Exp. Biol. Med.*, 1975, **148**, 194.
3 Lehmann, F.G., *Clin. Chim. Acta*, 1975, **65**, 257.
4 Schlaeger, R., *Z. Klin. Chem. Klin. Biochem.*, 1975, **13**, 277.
5 Nisselbaum, J.S., and Bodansky, O., *J. Biol. Chem.*, 1959, **234**, 3276.
6 Plagemann, P.G.W., Gregory, K.F., and Wroblewski, F., *J. Biol. Chem.*, 1960, **235**, 228.

7 Eppenberger, H.M., Dawson, D.M., and Kaplan, N.O., *J. Biol. Chem.*, 1967, **242**, 204.
8 Massarat, S., and Lang, N., *Klin. Wochenschr.*, 1965, **43**, 602.
9 Schlamowitz, M., and Bodansky, O., *J. Biol. Chem.*, 1959, **234**, 1433.
10 Boyer, S.H., *Ann. N.Y. Acad. Sci.*, 1963, **103**, 938.
11 Sussman, H.H., Small, P.A., and Cotlove, E., *J. Biol. Chem.*, 1968, **243**, 160.
12 Birkett, D.J., Done, J., Neale, F.C., and Posen, S., *Br. Med. J.*, 1966, **1**, 1210.
13 Usategui-Gomez, M., Yeager, F.M., and Fernandez de Castro, A., *Cancer Res.*, 1973, **33**, 1574.
14 Lehmann, F.G., *Clin. Chim. Acta*, 1975, **65**, 271.
15 Pfleiderer, G., Dikow, A.L., and Falkenberg, F., *Hoppe-Seyler's Z. Physiol. Chem.*, 1974, **355**, 233.
16 Boll, M., Backs, M., and Pfleiderer, G., *Hoppe-Seyler's Z. Physiol. Chem.*, 1974, **355**, 811.
17 Jockers-Wretou, E., and Pfleiderer, G., *Clin. Chim. Acta*, 1975, **58**, 223.
18 Neumeier, D., Knedel, M., Würzburg, U., Hennrich, N., and Lang, H., *Klin. Wochenschr.*, 1975, **53**, 329.
19 Neumeier, D., Prellwitz, W., Würzburg, U., Brundobler, M., Olbermann, M., Just, H.-J., Knedel, M., and Lang, H., *Clin. Chim. Acta*, 1976, **73**, 445.
20 Gerhardt, W., Ljungdahl, L., Börjesson, J., Hofvendahl, S., and Hedenäs, B., *Clin. Chim. Acta*, 1977, **78**, 29.
21 Lundin, A., and Styrélius, I., *Clin. Chim. Acta*, 1978, **87**, 199.
22 Fishman, W.H., Inglis, N.R., Green, S., Anstiss, C.L., Gosh, N.K., Reif, A.E., Rustigian, R., Krant, M.J., and Stolbach, L.L., *Nature*, 1968, **219**, 697.
23 Van Lente, F., and Galen, R.S., *Clin. Chim. Acta*, 1978, **87**, 211.
24 Uriel, J., *Ann. N.Y. Acad. Sci.*, 1963, **103**, 956.
25 Korngold, L., *Int. Arch. Allergy*, 1970, **37**, 366.
26 Foti, A.G., Herschman, H., and Cooper, J.F., *Cancer Res.*, 1975, **35**, 2446.
27 Markert, C.L., and Appella, E., *Ann. N.Y. Acad. Sci.*, 1963, **103**, 915.
28 Grimm, F.C., and Doherty, D.G., *J. Biol. Chem.*, 1961, **236**, 1980.
29 Kaminski, M., *Nature*, 1966, **209**, 723.
30 Moss, D.W., *Clin. Chim. Acta*, 1973, **43**, 447.
31 Jacoby, B., and Bagshawe, K.D., *Cancer Res.*, 1972, **32**, 2413.
32 Nørgaard-Pedersen, B., *Scand. J. Immunol.*, 1973, Suppl. 1, 125.
33 Laurell, C.-B., *Scand. J. Clin. Lab. Invest.*, 1972, Suppl. 124, 21.
34 Forman, D.T., Moss, D.W., and Whitaker, K.B., *Clin. Chim. Acta*, 1976, **68**, 287.
35 Milisauskas, V., and Rose, N.R., *Clin. Chem.*, 1972, **18**, 1529.
36 Kahn, A., Cottreau, D., Bernard, J.F., and Boivin, P., *Biomedicine*, 1975, **22**, 539.
37 Milisauskas, V., and Rose, N.R., *Exp. Cell Res.*, 1973, **81**, 279
38 Laurell, C.-B., *Anal. Biochem.*, 1965, **10**, 358.
39 Bøg-Hansen, T.C., and Daussant, J., *Anal. Biochem.*, 1974, **61**, 522.
40 Blomberg, F., and Raftell, M., *Eur. J. Biochem.*, 1974, **49**, 21.
41 Raftell, M., and Blomberg, F., *Eur. J. Biochem.*, 1974, **49**, 31.

42 Kohn, J., *J. Clin. Pathol.*, 1970, **23,** 732.
43 Kelkar, S.S., and Niphadkar, K.B., *Lancet*, 1974, **ii,** 1394.
44 Foti, A.G., Cooper, J.F., and Herschman, H., *Clin. Chem.*, 1978, **24,** 140.
45 Chu, T.M., Wang, M.C., Scott, W.W., Gibbons, R.P., Johnson, D.E., Schmidt, J.D., Leoning, S.A., Prout, G.R., and Murphy, G.P., *Invest. Urol.*, 1978, **15,** 319.
46 Vihko, P., Kontturi, M., and Korhonen, L.K., *Clin. Chem.*, 1978, **24,** 466.
47 Fang, V.S., Cho, H.-W., and Meltzer, H.Y., *Enzyme*, 1978, **23,** 210.
48 Headings, V.E., and Tashian, R.E., *Biochem. Genet.*, 1970, **4,** 285.
49 Geokas, M.C., Wollesen, F., Rinderknecht, H., and Martinson, F., *J. Lab. Clin. Med.*, 1974, **84,** 574.
50 Geiger, B., Navon, R., Ben-Yoseph, Y., and Arnon, R., *Eur. J. Biochem.*, 1975, **56,** 311.
51 Vihko, P., Sajanti, E., Janne, O., Peltonen, L., and Vihko, R., *Clin. Chem.*, 1978, **24,** 1915.
52 Iino, S., Abe, K., Oda, T., Suzuki, H., and Sugiura, M., *Clin. Chim. Acta*, 1972, **42,** 161.
53 Chang, C.-H., Raam, S., Angellis, D., Doellgast, G., and Fishman, W.H., *Cancer Res.*, 1975, **35,** 1706.
54 Holmgren, P.A., Stigbrand, T., Dumber, M.-G., and von Schoultz, B., *Clin. Chim. Acta*, 1978, **83,** 205.
55 Roberts, R., Sobel, B.E., and Parker, C.W., *Science*, 1976, **194,** 855.
56 Roberts, R., Parker, C.W., and Sobel, B.E., *Lancet*, 1977, **ii,** 319.
57 Willerson, J.T., Stone, M.J., Ting, R., Mukherjee, A., Gomez-Sanchez, C.E., Lewis, P., and Hersh, L.B., *Proc. Natl. Acad. Sci. U.S.A.*, 1977, **74,** 1711.
58 Van Steirteghem, A.C., Zweig, M.H., and Schechter, A.N., *Clin. Chem.*, 1978, **24,** 414.
59 Zweig, M.H., Van Steirteghem, A.C., and Schechter, A.N., *Clin. Chem.*, 1978, **24,** 422.
60 Tedesco, T.A., and Mellman, W.J., *Science*, 1971, **172,** 727.
61 Tedesco, T.A., *J. Biol. Chem.*, 1972, **247,** 6631.
62 Goedde, H.W., and Altland, K., *Ann. N.Y. Acad. Sci.*, 1968, **151,** 540.
63 Rubinstein, H.M., Dietz, A.A., Lubrano, T., and Garry, P.J., *J. Med. Genet.*, 1976, **13,** 43.
64 Rubinstein, H.M., Dietz, A.A., and Lubrano, T., *J. Med. Genet.*, 1978, **15,** 27.
65 Rubin, C.S., Dancis, J., Yip, L.C., Nowinsky, R.C., and Balis, M.E., *Proc. Natl. Acad. Sci. U.S.A.*, 1971, **68,** 1461.
66 Arnold, W.J., Meade, J.C., and Kelly, W.N., *J. Clin. Invest.*, 1972, **51,** 1805.
67 Dreyfus, J.C., and Alexandre, Y., *Biochem. Biophys. Res. Commun.*, 1971, **44,** 1364.
68 Shibuta, Y., Higashi, T., Hirai, H., and Hamilton, H.B., *Arch. Biochem. Biophys.*, 1967, **118,** 200.
69 Shapira, E., Ben-Yoseph, Y., Eyal, F.G., and Russell, A., *J. Clin. Invest.*, 1974, **53,** 59.
70 Shapira, F., Nordmann, Y., and Gregori, C., *Acta Med. Scand.*, 1972, Suppl. 542, 77.

71 Stumpf, D., Neuwelt, E., Austin, J., and Kohler, P., *Arch. Neurol.*, 1971, **25**, 427.
72 Cox, R.P., Elson, N.A., Tu, S., and Griffin, M.J., *J. Mol. Biol.*, 1971, **58**, 197.
73 Levitt, M.D., and Cooperband, S.R., *N. Engl. J. Med.*, 1968, **278**, 474.
74 Kanno, T., and Sudo, K., *Clin. Chim. Acta*, 1977, **76**, 67.
75 Kobayashi, T., Nakayama, T., and Kitamura, M., *Clin. Chim. Acta*, 1978, **86**, 261.
76 Thomas, D.W., Rosen, S.W., Kahn, R., Temple, R., and Papadopoulos, N.M., *Ann. Intern. Med.*, 1974, **81**, 434.
77 Biewenga, J., *Clin. Chim. Acta*, 1977, **76**, 149.
78 Nagamine, M., and Ohkuma, S., *Clin. Chim. Acta*, 1975, **65**, 39.
79 Crofton, P.M., and Smith, A.F., *Clin. Chim. Acta*, 1978, **83**, 235.
80 Lee, C., Wang, M.C., Murphy, G.P., and Chu, T.M., *Cancer Res.*, 1978, **38**, 2871.

5

Catalytic Differences between Multiple Forms of Enzymes

Although, by definition, the multiple forms of a particular enzyme all catalyse the same chemical reaction, they are not necessarily identical in their catalytic properties; indeed, variations between isoenzymes in these properties appear to be the rule rather than the exception. Differences in these respects, if expressed *in vivo*, have an obvious relevance to the physiological or pathological significance of enzyme variation, while *in vitro* they provide an important set of criteria for isoenzyme characterisation.

Differences in Specific Activity

Reduced catalytic activity of the affected isoenzymes is a frequent accompaniment of allelic variation, increased activity compared with the product of the usual allele being much less common. It is the reduction in enzymic activity associated with some rare alleles which accounts for their pathological significance as the carriers of inherited disease. Differences in catalytic activity associated with variant isoenzymes may result from several factors, besides a change in catalytic effectiveness of the enzyme molecule; *e.g.*, a reduced rate of synthesis, or greater lability and reduced half-life, of a mutant enzyme may account for its lower activity *in vivo*. These alternatives cannot readily be distinguished, unless an independent means of estimating the amount of protein synthesised by the allelic gene is available. This can be achieved by immunochemical titration in those cases in which the gene product retains its antigenic identity. A greater than normal rate of synthesis of one variant of glucose-6-phosphate dehydrogenase, G6PD Hektoen, has been demonstrated, as well as reduced rates of synthesis of other variants. In at least one condition, a variant form of the inherited disease galactosaemia, an isoenzyme of the affected enzyme, galactose-1-phosphate uridylyl transferase (EC 2.7.7.12; also called hexose-1-phosphate uridylyl transferase) is synthesised in normal quantities but has only half the usual specific activity.[1] Other examples of the production by mutant genes of isoenzymes with altered specific activities are known, and undoubtedly this explanation will be found to account for many genetically determined variations in enzyme activity. However, because of the difficulties of devising specific techniques for determination of the amount of enzyme protein or of obtaining isoenzymes in a sufficiently pure form for the direct measurement of specific activity, differences in this property are of limited value in isoenzyme characterisation.

Differences in Reaction with Substrates

In Michaelis Curves.—The dependence of the velocity of an enzymic reaction on substrate concentration is expressed for the majority of enzymes by the Michaelis–Menten equation, $v = V . s/(s + K_m)$, in which v is the initial rate of reaction, s the concentration of substrate, V the maximum velocity and K_m the Michaelis constant. Exceptions to the hyperbolic relationship between v and s described by the equation occur with the important category of allosteric enzymes, for which a sigmoid relationship is observed under certain conditions (*i.e.*, the Michaelis–Menten equation does not hold at low substrate concentrations), while for many enzymes deviations from Michaelis–Menten kinetics are seen at high substrate concentrations (inhibition by excess of substrate). Since, also, the majority of enzymic reactions involve more than one substrate, the dependence of v on the concentration of one substrate is influenced by that of the second substrate in a manner related to the reaction mechanism. Nevertheless, studies of velocity–substrate relationships have fulfilled an important role in isoenzyme studies, both in the identification and characterisation of multiple enzyme forms and in attempting to infer their possible physiological significance.

The value of V calculated from a series of measurements of v at different substrate concentrations is directly proportional to the amount of enzyme added; since purified isoenzymes are rarely available, V is therefore not a particularly useful basis for comparison of isoenzyme preparations. On the other hand, K_m, the substrate concentration at which the observed reaction velocity is half V, is usually independent of enzyme concentration (provided, of course, that a fixed concentration is maintained in a single experiment) and, although K_m may be affected by several other variables, this parameter is a valuable means of denoting differences between isoenzymes in affinity for their substrates. Michaelis constants are usually dependent on pH, and may be influenced by other factors such as temperature, the nature of buffer ions, ionic strength, the presence and concentration of cofactors and, as already mentioned, the concentration of a second substrate in some two-substrate reactions. Careful standardisation of experimental conditions is therefore essential for reliable comparisons of K_m values of isoenzymes, especially as the values may differ only slightly. The range of substrate concentrations should extend above and below the K_m value. Since v is low at small concentrations of s and its measurement therefore subject to greater experimental error, appropriate weighting should be given to individual values of v to allow for this when calculating the Michaelis constant.

Differences in Michaelis constants have been reported for the members of several sets of isoenzymes determined by allelic genes.[2] Although less-common isoenzymes generally show a reduced affinity for their substrates, *i.e.*, increased K_m values, compared with the more usual form, variants of glucose-6-phosphate dehydrogenase have been described which have increased substrate-affinity, as well as forms which show decreased substrate-binding. The range of variation of K_m values for glucose-6-phosphate in this family of mutant enzymes is of the order of one-third to four times the value for the common isoenzyme. Michaelis constants for the coenzyme, NADP, also differ between some variants. Isoenzymes with reduced affinity are less efficient catalysts at low substrate concentrations and this presumably contributes to their relative functional ineffectiveness in the cells in which they occur.

Isoenzymes determined by multiple gene loci also typically differ in their Michaelis constants. The value for the H_4 tetramer of human lactate dehydrogenase is approximately one-tenth of that for the M_4 isoenzyme with pyruvate as substrate, and rather less than half in the reverse reaction in which the substrate is lactate. The heteropolymers have intermediate Michaelis characteristics. The LD_1 isoenzyme is inhibited by excess of pyruvate to a greater extent than LD_5 when measurements are made under certain conditions.[3] A theory of the physiological functions of the LD_1 and LD_5 isoenzymes was based originally on this difference in properties, since the ability of tissues to accumulate pyruvate and to convert it to lactate under anaerobic conditions is well correlated with the concentrations within them of the less-inhibited LD_5 isoenzyme.[4] However, the properties of the enzymes in dilute solution do not accurately reflect those in living tissues; furthermore, concentrations of pyruvate in muscle do not reach inhibitory levels, even after severe anaerobic exercise.[5]

Other examples of isoenzymic products of multiple gene loci which exhibit differences in Michaelis constants include the MM and BB forms of creatine kinase[6, 7, 8] and the A, B, and C isoenzymes of aldolase,[9] as well as the genetically distinct cytoplasmic and mitochondrial isoenzymes of such enzymes as aspartate aminotransferase[10, 11, 12] and malate dehydrogenase.[13] As well as differing in K_m values, the isoenzymes may also differ in inhibition by excess of substrate or by a product of the reaction; *e.g.*, mitochondrial aspartate aminotransferase is more susceptible to inhibition by oxaloacetate accumulating during the course of the reaction than the cytoplasmic isoenzyme, with the direction of reaction usually chosen for measurement of aminotransferase activity.[11] Since oxaloacetate is also a substrate of the reverse reaction, this corresponds to a greater inhibition of the mitochondrial isoenzyme by an excess of this substrate when the reaction is reversed. When hybrid isoenzymes composed of aggregates of unlike subunits can exist, they possess kinetic properties intermediate between those of the homopolymers, as do the heteropolymers of lactate dehydrogenase.

Similarity of Michaelis constants has been used to support the view that some multiple forms of enzymes separable by charge-dependent means do not represent distinct isoenzymes, but are more probably attributable to such phenomena as aggregation of a single type of enzyme or binding of enzyme molecules to non-enzymic proteins. This approach was employed in the case of the minor zones of alkaline phosphatase seen after starch-gel electrophoresis of tissue extracts (Fig. 1.6). Michaelis constants for this enzyme are markedly pH dependent and inter-tissue differences in K_m values are small. However, by

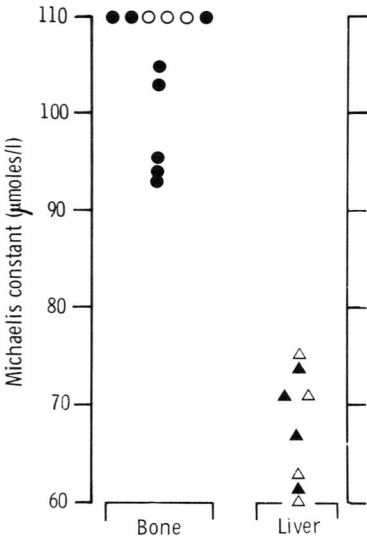

Figure 5.1 *Michaelis constants of alkaline phosphatase isolated by starch-gel electrophoresis from sera of patients with liver or bone diseases (solid symbols), compared with values obtained by adding extracts of human bone or liver to serum and recovering the respective tissue phosphatases in the same way (open symbols). The substrate was β-naphthyl phosphate and reaction velocities were measured at optimum pH at each substrate concentration.* (Data from reference 16)

measuring velocities at the optimum pH value for each substrate concentration, reproducible differences in K_m between alkaline phosphatases from several human tissues were demonstrated,[14] differences which were also displayed by the minor enzyme components of each tissue.[15] The small differences between liver and bone in the Michaelis constants of the major components of their respective alkaline phosphatases were found to be reflected in alkaline phosphatases extracted by starch-gel electrophoresis from the sera of patients with either hepatobiliary or bone diseases,[16] providing evidence of the tissues of origin of the raised serum alkaline phosphatase activity in these patients (Fig. 5.1).

When two or more isoenzymes with different Michaelis constants act on a single substrate, non-linear plots are obtained with the various transformations of the Michaelis–Menten equation (*e.g.*, the double-reciprocal plot of $1/v$ against $1/s$) which typically yield straight lines with single enzymes. For instance, plots of $1/v$ against $1/s$ are convex upwards, the curvature becoming more marked at higher substrate concentrations whereas, at lower concentrations, the line straightens, tending to cut the abscissa at a value approaching $-1/K_m$ for the isoenzyme with the smallest Michaelis constant. This is because, at lower substrate concentrations, the enzyme with the greatest affinity for the substrate (*i.e.*, the smallest K_m value) preferentially binds substrate molecules. Non-rectilinear plots of data according to the transformed Michaelis–Menten equation are therefore a useful indication of heterogeneity of enzyme preparations acting on a single substrate and the possible existence of kinetically distinct isoenzymes. However, when differences in K_m values are small and if a relatively limited range of substrate concentrations is used, deviation of the plot from a straight line may be imperceptible, giving an apparent Michaelis constant intermediate between the extreme values for the isoenzymes composing the mixture. Apparently rectilinear plots of this type were obtained for mixtures of liver and bone alkaline phosphatases in serum.[16]

Differences in kinetic properties between isoenzymes are significant in devising conditions for the assay of total enzymic activity in mixtures of isoenzymes. On the other hand, numerous attempts have also been made to exploit these differences to estimate the respective contributions of individual isoenzymes in such mixtures.[3, 17, 18] Because of the difference in Michaelis constants of the LD_1 and LD_5 isoenzymes of lactate dehydrogenase the ratio of activities at high and low concentrations of pyruvate will depend on the relative proportions of these two isoenzymes in the enzyme sample. With pyruvate concentrations of 1.2 and 0.15 mmoles l^{-1} pyruvate, a ratio of approximately 0.4 is observed with extracts of rabbit tissues such as heart or erythrocytes rich in LD_1, compared with 1.5—2.0 for liver or skeletal-muscle extracts.[3] Differences in inhibition by excess of substrate of lactate dehydrogenase in extracts of human heart or liver tissue are such that the activity of the heart enzyme is reduced by approximately 60% by increasing the pyruvate concentration from 0.34 to 5 mmoles l^{-1}, whereas that of the liver enzyme falls by only 18%.[18] Although ratios of activities at two substrate concentrations broadly reflect the relative preponderance of the H and M subunits in enzyme preparations, clearly distinct ratios are not observed when enzyme samples such as serum are analysed, since these may differ to a comparatively small extent in the proportions of heteropolymeric isoenzymes which they contain, as well as in their content of the H_4 and M_4 homopolymers. Determination of activity-ratios at two concentrations of pyruvate is therefore an insensitive analytical procedure for multi-component lactate-dehydrogenase isoen-zyme systems.

Analysis of isoenzyme mixtures by differences in substrate affinities is much simplified when only two components are present. The relative proportions of the cytoplasmic and mitochondrial isoenzymes of aspartate aminotransferase in tissue extracts have been estimated by measuring activity with low aspartate and high 2-oxoglutarate con-centrations, conditions which favour the mitochondrial isoenzyme, whereas at high concentrations of both substrates activity is due to the two isoenzymes.[10] In this method, discrimination is aided by determining the activity at low aspartate concentration at pH 6.0, rather than at pH 7.4 as is usual at high concentrations, thus taking advantage of differences in pH dependence of the mitochondrial and cytoplasmic isoenzymes.

When the object of determining creatine kinase isoenzymes in serum is the assessment of myocardial damage, absence of the BB dimer can usually be assumed so that the problem is reduced to the measurement of the activity of MB-creatine kinase in the presence of an excess of the MM form. The MB isoenzyme has a greater affinity for creatine phosphate than has the MM dimer, the ratio of K_m values for the purifed isoenzymes approaching 1:3. There is also a slight difference between the K_m values for the second substrate, adenosine diphosphate.[8] The effect of these differences on the dependence of v on s has been used to measure the proportion of MB-creatine kinase in serum samples: activity is determined at a creatine phosphate concentration of 1.2 mmoles l^{-1} and with 18 mmoles l^{-1} of adenosine diphosphate, then with 2.5 mmoles l^{-1} and 12 μmoles l^{-1}, respectively, of these two substrates.[7] The ratio of activities under the two sets of conditions was found to be related rectilinearly to the percentage of MB-creatine kinase in mixtures of the MM and MB isoenzymes. A particularly sensitive assay is needed to measure creatine kinase activity at the extremely low second con-centration of adenosine diphosphate. The necessary sensitivity of measurement of ATP production by creatine kinase can be achieved by the use of a coupled assay system, in which light is emitted in the presence of luciferin and firefly luciferase.[7] However, the requirement for an especially sensitive enzyme assay, with the increased variability of estimates of low catalytic activities, emphasises the difficulties of methods of isoenzyme analysis which depend on the use of very low substrate concentrations. Experimental errors are further magnified by the need to calculate results in terms of ratios of two separate activity measurements.

With Substrate Analogues.—The degree of specificity of enzymes for their substrates shows a wide variation, from absolute specificity at one extreme, in which the enzyme is completely inactive towards all compounds other than a single, unique substrate, to examples in which the only requirements for catalysis are the presence of a particular chemical grouping or type of bond in the putative substrate molecules. Substrate analogues obviously cannot be used in the study of isoenzymes with absolute specificity. However, when substrate specificity is less than absolute, isoenzymes frequently differ considerably, both quantitatively and qualitatively, in their reactivity towards substrate analogues. Differences are most marked among members of families of enzymes with group- or bond-specificity, such as the non-specific phosphatases or carboxylate esterases. Indeed, a distinction between multiple forms of a single enzyme and a group of individual enzymes with overlapping substrate specificities is difficult to draw, and cannot be based solely on relative rates of hydrolysis of a range of synthetic esters. Genetic studies may establish whether allelic variation or multiple gene loci are responsible for functionally similar esterases. Although demonstration of allelic variation clearly defines the gene-products as isoenzymes, the inclusion of multiple forms of esterases determined by multiple gene loci within the category of isoenzymes will remain a matter of preference, depending on the weight given to known functional and structural similarities. Nevertheless, the characterisation of the many multiple forms of non-specific carboxylic-esterases in human and other tissues has depended heavily on electrophoretic separation of enzyme zones, followed by comparison of the patterns obtained with various synthetic ester substrates, and this constitutes one of the earliest examples of the use of this approach to the study of multiple enzyme forms.[19] In the case of ali-esterases, for example, alternative substrates consist typically of esters with acyl components of various chain-lengths, esterified with α- or β-naphthols, liberation of which by hydrolysis can readily be detected by fluorescence, or by coupling with diazonium salts to produce coloured dyes. Derivatives of indoxyl can be used in a similar manner, air-oxidation of indoxyl to indigo eliminating the need for a further chromogenic reaction.[20] Changes in the relative intensities of the various enzyme zones with different substrates can be used to assess substrate specificity. More reliable quantitative estimates of the relative reactivity of esterase preparations towards various substrates can be obtained by measuring rates of hydrolysis in solution.

The various multiple forms of non-specific acid and alkaline phosphatases are amenable to characterisation by the use of alternative substrates, since almost all orthophosphate esters are hydrolysed by these enzymes. In the case of human alkaline phosphatases, the only structural requirements for a potential substrate are the presence of a terminal orthophosphoric acid radical, two hydroxyl groups of which are unesterified; thus, inorganic pyrophosphate or polyphosphates such as ADP or ATP are cleaved by alkaline phosphatases, orthophosphate groups being removed sequentially from polyphosphate substrates.[21] Consequently, many studies have been made of the relative activities of both acid and alkaline phosphatases towards various derivatives of orthophosphoric acid.

Among acid phosphatases from various tissues, differences in the relative rates of hydrolysis of α- and β-glycerophosphates by the enzymes from erythrocytes and spleen were demonstrated as early as 1934, and further differences with respect to these substrates and phenyl phosphate were later shown for prostatic acid phosphatases as well as for the other two enzymes.[22, 23] Investigations of the substrate specificity of human acid phosphatases have continued to the present day, because of the clinical requirement for methods of assay with high specificity for the prostatic enzyme, and additional phosphate esters studied include p-nitrophenyl phosphate, α- and β-naphthyl phosphates, and phenolphthalein and thymolphthalein monophosphates.[24, 25]

Differences in relative rates of hydrolysis of alternative substrates by tissue-specific forms of alkaline phosphatase are well established. As with many other criteria which have been applied to this group of enzymes from human and other animal tissues, relative activities with various substrates tend to separate the alkaline phosphatases into two main categories, one composed of the isoenzymes from such tissues as bone, liver, or kidney, the other consisting of the placental and small-intestinal phosphatases. Thus, with β-glycerophosphate as a reference substrate, p-nitrophenyl phosphate is a relatively poor substrate for the human intestinal isoenzyme while adenosine monophosphate is rapidly hydrolysed, a reversal of the patterns of activity found for the enzymes of liver, kidney, and bone.[26] Placental alkaline phosphatase is relatively more active towards β-glycerophosphate compared with phenyl phosphate than kidney phosphatase, though with this pair of substrates the intestinal and kidney isoenzymes are rather similar in their relative activities.[27] Dissimilarities in substrate specificity also extend to

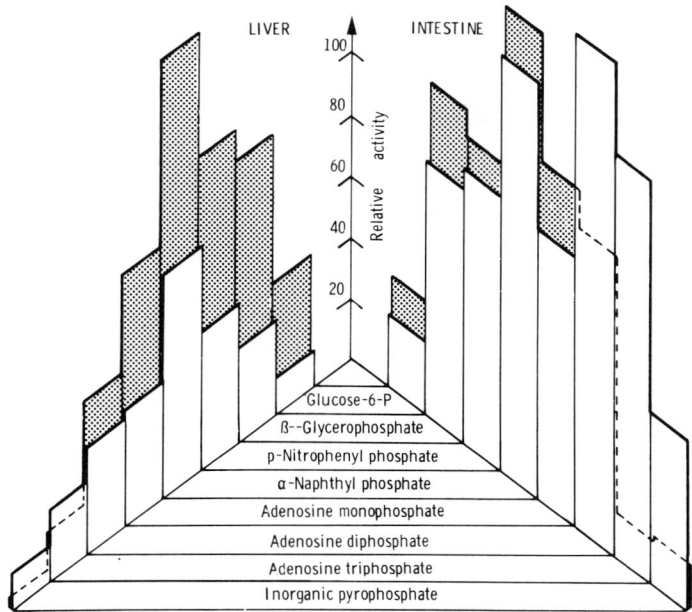

Figure 5.2 *Relative rates of reaction of alkaline phosphatases from human liver and small intestine with a variety of substrates at pH 9.5, with (shaded bars) and without (open bars) added magnesium ions. The activities for each isoenzyme have been related to a value of 100 for hydrolysis of p-nitrophenyl phosphate in the presence of magnesium.*

pyrophosphate substrates such as inorganic pyrophosphate, ADP and ATP, with relatively more rapid hydrolysis of these compounds being effected by intestinal alkaline phosphatase than by non-intestinal isoenzymes when phenyl orthophosphate or a derivative of it is the reference substrate[28, 29] (Fig. 5.2).

An extensive investigation of the substrate specificities of alkaline phosphatases from rat tissues also demonstrated marked differences between intestinal and other phosphatases in this respect, particularly with regard to the relative sensitivity of o-carboxyphenyl phosphate to hydrolysis by the intestinal enzyme, compared with the resistance of this substrate to attack by the enzyme from liver.[30] This work indicated the possibility of quantitative analysis of mixtures of intestinal and non-intestinal alkaline phosphatases by the measurement of relative rates of hydrolysis with two substrates, a principle which was subsequently applied to human serum.[31] The latter study confirmed the suitability of o-carboxyphenyl phosphate compared with phenyl phosphate as a

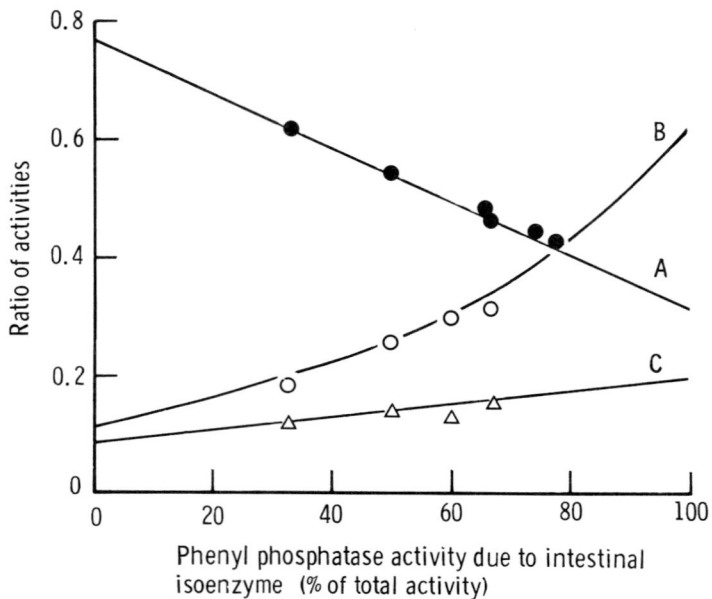

Figure 5.3 *Ratios of rates of hydrolysis of three pairs of substrates by mixtures of alkaline phosphatases in human serum as a function of the proportion of the intestinal isoenzyme present. The latter is expressed as a percentage of the total phenyl phosphatase activity. Lines A, B, and C were calculated for the substrate pairs thymolphthalein and phenyl phosphates, o-carboxyphenyl and thymolphthalein phosphates, and o-carboxyphenyl and phenyl phosphates, respectively, on the basis of activity ratios measured with preparations of liver and intestinal isoenzymes. The symbols indicate ratios determined experimentally for mixtures of these two phosphatases in serum.* [Reproduced, with permission, from: D. W. Moss, *Clin. Chim. Acta* (Elsevier), 1971, **35**, 413]

substrate for human intestinal phosphatase. However, the rather low rate of hydrolysis of this substrate, even by intestinal phosphatase, limits the accuracy of measurements of its breakdown; therefore, thymolphthalein phosphate : phenyl phosphatase ratios were preferred in deriving the relative contributions of intestinal and non-intestinal isoenzymes to the total alkaline phosphatase activity of serum, lower ratios being observed with increasing proportions of the intestinal enzyme (Fig. 5.3).

Although the alkaline phosphatases are hydrolases, hydroxy-compounds other than water can act as acceptors of the phosphate group from the primary substrate when present in high concentrations. Some human isoenzymes differ to a certain extent in the relative rates at which the phosphate radical is transferred to such compounds as tris(hydroxymethyl)aminomethane, diethanolamine, or 2-amino-2-methyl-1-propanol.[32, 33] However, these differences have not yet proved useful in isoenzyme characterisation.

The isoenzymes of lactate dehydrogenase differ in their relative activities towards higher homologues of their natural substrate, L(+)-lactic acid, and this property has also been exploited as an aid to clinical diagnosis. The hydroxy-derivatives of butyric, caproic, and valeric acids act as substrates for these isoenzymes, as do the corresponding oxo-compounds in the reverse reaction. Since the reverse reaction is the more rapid, pyruvate and analogues of it are usually chosen in analytical methods based on the differential substrate specificities of lactate dehydrogenase isoenzymes. The ratios of activities with 2-oxobutyrate as substrate to those with pyruvate are of the order of 1 for the most-anodal isoenzyme, LD_1, from human tissues, but less than 0.2 for LD_5, with concentrations of the two substrates of 3.3 and 0.7 mmoles l^{-1}, respectively, and when activities are measured at 25 °C.[34, 35] Thus, sera in which total lactate dehydrogenase is raised due to enzyme release from a tissue rich in LD_1, such as heart muscle, have higher 2-oxobutyrate:pyruvate activity ratios than is the case when LD_5 is released, *e.g.*, as a result of hepatitis. The method has been widely applied in clinical analysis, particularly to give increased sensitivity of detection of myocardial damage.[36, 37] However, it suffers from the general disadvantages of methods of this type, in that it is insensitive to alterations in the proportions of the intermediate isoenzymes, while the need to determine ratios of two separate measurements increases analytical error. Furthermore, the ratios characteristic of the fast and slow isoenzymes are dependent upon the reaction conditions so that the differences between the isoenzymes may be reduced if these are altered.

Isoenzymes of dehydrogenases are also active to varying degrees when the naturally occurring second substrates of the reactions which they catalyse, nicotinamide-adenine dinucleotide (NAD) or nicotinamide-adenine dinucleotide phosphate (NADP), are replaced by various synthetic analogues of these coenzymes. The effects of various substituent groups in either the pyridine or the purine rings have been investigated.[38] As with experiments in which the first substrate of the reaction is varied, the results of replacing the natural coenzyme with various synthetic analogues are usually expressed as ratios of the activities observed with pairs of alternative coenzymes. Differential effects of coenzyme modifications on the activities of lactate dehydrogenase isoenzymes of several species have been demonstrated.[39] With either the 3-thionicotinamide or 3-acetylpyridine analogues as hydrogen-acceptors, for example, ratios for the oxidation of lactate by extracts of heart muscle (*i.e.*, a tissue rich in LD_1) from birds, mammals, amphibia, and fish all fall within the range of 5—8. Values for skeletal muscle are much more heterogeneous, ranging from as low as 0.2 in fish to more than 2 in man, reflecting the variable isoenzyme composition of this tissue from one species to another. As with many other catalytic properties of lactate dehydrogenases, the heteropolymeric isoenzymes show a graded reactivity towards substrate analogues, between the extremes represented by the respective homopolymers.

Other examples of the use of coenzyme analogues to characterise the isoenzymes of dehydrogenases include the differentiation of the cytoplasmic and mitochondrial malate dehydrogenases of ox heart and rabbit muscle, and isoenzymes of this enzyme from various tissues of snail, clam, and octopus.[13,38] Distinct forms of dehydrogenase acting on compounds which contain vicinal hydroxyl groups, such as glycerol, are produced by *Aerobacter aerogenes* depending on whether the organism is grown on glycerol- or glucose-containing media. These enzyme forms are not physically separable, but can be distinguished by their reactivity with analogues of NAD.[38] Malate dehydrogenase also resembles lactate dehydrogenase in that some variation in the structure of the non-coenzyme substrate of the reaction can be tolerated; differentiation between cytoplasmic and mitochondrial isoenzymes, and between homologous isoenzymes from various rat tissues, has been achieved by measuring rates of reduction of mono- and di-fluoro-derivatives of oxaloacetate in the reverse reaction catalysed by the enzyme.[40]

Selective Inhibition of Isoenzymes

Variations between isoenzymes in their Michaelis constants for particular substrates can be interpreted as reflecting minor differences in the three-dimensional structures of the active centres at which binding of substrate molecules takes place. It is not surprising, therefore, that such structural variations should also cause isoenzymes to differ in their affinities for, and responses to, specific inhibitors, not only when these substances bind to the active centre itself as is the case for competitive inhibitors, but also when other specialised regions of the enzyme molecule are involved in the attachment of the inhibitor. Specific inhibitory effects of this nature are to be distinguished from the irreversible, relatively non-specific inactivation processes described in Chapter 3. Specific enzyme inhibition is typically not time dependent and is reversible, *i.e.*, removal of the inhibitor by a physical separation technique, such as dialysis, restores catalytic activity. Some cases of specific enzyme inhibition involve covalent attachment of the inhibitor to the enzyme and therefore are time dependent and irreversible by physical separation, although chemical means of breaking the enzyme-inhibitor bond may re-activate the enzyme.

Early uses of specific inhibitors to establish organ-specific characteristics of enzymes included attempts to discriminate between prostatic and non-prostatic forms of non-specific acid phosphatase, in order to increase the diagnostic value of acid-phosphatase measurements in serum in the detection of disease of the prostate. Some selective inhibitors of acid phosphatases, such as formaldehyde and organic solvents, which probably act by denaturation of the enzyme molecules, have been mentioned in Chapter 3. However, other compounds are reversible inhibitors and, of these, the most useful is dextro-rotatory tartrate. Almost complete inhibition of the prostatic isoenzyme is observed at concentrations of tartrate which have no effect on the red-cell enzyme.[23] Although prostate is not the only tissue which contains tartrate-inhibited acid phosphatase, elevation of the activity of this isoenzyme in serum is essentially confined to cases of metastatic carcinoma of this tissue, so that the clinical value of this inhibitor is considerable. Dextro-rotatory tartrate is a stereospecific, fully competitive inhibitor of the prostatic isoenzyme, with an inhibitor constant of approximately 0.04 mmoles l^{-1}; *i.e.*, of an order of magnitude similar to that of the Michaelis constants for commonly used substrates (about 0.1 mmoles l^{-1}).[41] Since the degree of inhibition by a competitive inhibitor depends on the ratio of the affinity constants of substrate and inhibitor and their relative concentrations, a tartrate concentration of the order of 50 mmoles l^{-1} is usually chosen to ensure virtually complete inhibition of the prostatic isoenzyme in assays with substrate concentrations of 5 mmoles l^{-1}.

Several inhibitors have been shown to exert differential effects on the various isoenzymes of lactate dehydrogenase. The most-anodal isoenzyme, LD_1, is inhibited by sulphite to a greater extent than is the case for LD_5; with a concentration of the inhibitor of 2×10^{-2} mmoles l^{-1}, these two isoenzymes from rat tissues are inhibited to the extent of about 70% and 30%, respectively.[42, 43] Lactate dehydrogenases are also inhibited by oxamate and oxalate. Oxamate is a competitive inhibitor of the reduction of pyruvate and a non-competitive inhibitor of the oxidation of lactate, whereas, with oxalate, the modes of inhibition are reversed. Both these inhibitors produce relatively greater effects on the activities of the more-anodal isoenzymes compared with their effects on the more basic enzyme forms.[44]

An early indication of the existence of at least two catalytically non-identical groups of alkaline phosphatases in mammalian tissues was provided by the less pronounced inhibition of intestinal alkaline phosphatase by bile acids than of phosphatases from other sources.[45] Further confirmation has come from observations with several compounds which inhibit alkaline phosphatases uncompetitively. This type of inhibition, which is uncommon in single-substrate reactions, occurs by combination of the inhibitor with the enzyme-substrate complex. A plot of $1/v$ against $1/s$ in the presence of an uncompetitive inhibitor is a straight line, parallel to that obtained when the inhibitor is absent; *i.e.*, both V_{max} and the apparent K_m are reduced by such an inhibitor. Amino acids have long been recognised as potential inhibitors of alkaline phosphatases, some exerting their effects in a non-specific manner which is probably related to interactions with activating or constituent metal ions. However, organ-specific, uncompetitive inhibition was first demonstrated with L-phenylalanine and later with other amino acids.[30,46,47,48] Compounds such as L-phenylalanine and L-tryptophan are more inhibitory towards the isoenzymes from placenta, small intestine, and some tumours than those from other tissues, whereas the reverse is the case for L-arginine and more particularly for L-homoarginine, a potent inhibitor of alkaline phosphatases from bone or liver. L-Leucine specifically inhibits certain rare variants of placental alkaline phosphatase, as well as forms of the enzyme occasionally detectable in sera of cancer patients. In all these cases inhibition is stereospecific, the D-isomer of the amino acid being inactive. The broad-spectrum anti-helminthic, levamisole (tetramisole; [(−)-2,3,5,6-tetrahydro-6-phenylimidazo(2,1-*b*) thiazole hydrochloride] is also an uncompetitive inhibitor of alkaline phosphatases from human and animal tissues; in this case isoenzymes from tissues other than placenta or small intestine such as bone, liver, or kidney, are the more affected.[49,50] The derivative of levamisole known as R8231 [(±)-6(*m*-bromophenyl)-5,6-dihydroimidazo(2,1-*b*) thiazole oxalate] is an even more potent inhibitor with similar specificity.

None of these inhibitors of alkaline phosphatases is completely specific for the isoenzyme from a particular tissue or group of tissues, nor is inhibition of the more sensitive isoenzymes complete, although activity may be almost entirely abolished by the more effective inhibitors such as levamisole. The extent of inhibition depends on both inhibitor and substrate concentrations, since the inhibitor combines with the enzyme–substrate complex. With a typical substrate concentration of 5 mmoles l^{-1}, L-phenylalanine at a similar concentration inhibits human intestinal alkaline phosphatase by about 70%, compared with 10% for the liver isoenzyme. Levamisole produces an equivalent inhibition of liver or bone alkaline phosphatases in concentrations of 0.05 mmoles l^{-1} with little effect on the placental or intestinal isoenzymes, while its analogue R8231 is equally effective at about one tenth of that concentration. The use of controls containing D-phenylalanine has been recommended to correct for non-specific inhibition or activation by amino acids when L-phenylalanine is used as a selective inhibitor of intestinal alkaline phosphatase, *e.g.*, in the detection and measurement of this isoenzyme in human serum.[51,52]

Specific inhibitors have played a particularly significant part in the detection and characterisation of several forms of serum cholinesterase with reduced catalytic activity determined by rare allelic genes. The isoenzyme produced by the so-called 'atypical' allele is less susceptible to inhibitors which contain a positively-charged nitrogen atom than the more common ('usual') form of the enzyme. Several compounds are inhibitors of serum cholinesterase, but the most generally used is dibucaine.[53] The percentage inhibition at a dibucaine concentration of 10^{-5} moles l^{-1} (the 'dibucaine number') is 80 ± 2 for the typical enzyme and 20 ± 1 for the atypical variant. Differences in the catalytic properties of these two isoenzymes also extend to their Michaelis constants for various substrates, values of these being significantly higher in the case of the atypical isoenzyme. Mutation is therefore thought to have produced an alteration at the site at which the charged group of the substrate or inhibitor is bound, so that affinity for both classes of reagents is reduced.

The usual isoenzyme is inhibited by fluoride and so, to a lesser extent, is the atypical variant. However, the existence of a third, catalytically

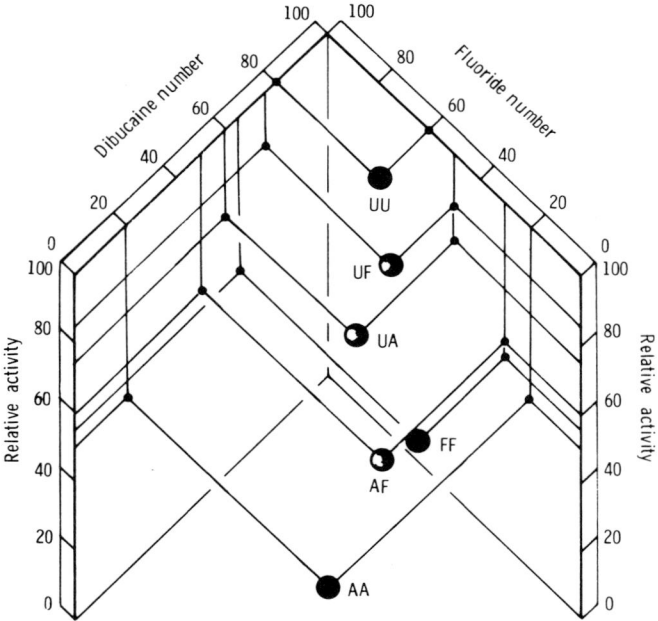

Figure 5.4 *Some catalytic characteristics of cholinesterase activity in the sera of individuals homozygous for the 'usual' gene (UU), the 'atypical' gene (AA) and the 'fluoride resistant' gene (FF). Heterozygotes for these three genes (UA, UF, and AF) have a mixture of isoenzymes in their sera with resultant catalytic properties intermediate between those of the respective homozygotes*

distinct isoenzyme was deduced from the observation that fluoride-resistance and dibucaine-resistance were not invariably correlated. The existence of a 'fluoride resistant' isoenzyme was therefore postulated.[54] When inhibition by fluoride at a concentration of 5×10^{-5} moles l^{-1} is expressed as a 'fluoride number', analogous to the dibucaine number, values of the order of 60 are observed for the usual isoenzyme, compared with 20 and 30, respectively, for the atypical and fluoride-resistant variants. The dibucaine number of the fluoride-resistant isoenzyme is about 65. Thus, a combination of measurements of inhibition by dibucaine and by fluoride is needed fully to characterise the two variant isoenzymes (Fig. 5.4). Individuals heterozygous for the genes determining the usual, atypical or fluoride-resistant isoenzymes have in their sera mixtures of various pairs of these isoenzymes; therefore, their serum cholinesterases show properties intermediate between those of the corresponding homozygotes, in inhibition characteristics as well as in total enzymic activity.

Differences between Isoenzymes in Other Catalytic Properties

Enzyme activity is frequently enhanced by the presence in the reaction mixture of other substances besides the enzyme and its substrate. Some of these activators are relatively unspecific in their actions; *e.g.*, various divalent metal ions activate a wide variety of different enzymic reactions. The mechanism of action of such activators is not known in many cases but a metal ion may act as an electropositive centre at the active site of the enzyme which assists substrate-binding, or the ion and the substrate molecule may form a complex which is the true substrate of the reaction, or the enzyme molecule may assume a catalytically more-favourable conformation in the presence of the ion. The reaction velocity typically shows a hyperbolic, Michaelis–Menten type of dependence on the concentration of activator, and isoenzymes may differ somewhat in their 'activator constants', analogous to the Michaelis constant, derived from this relationship. However, such differences are often small and therefore not useful in isoenzyme characterisation.

Some activators produce their effects by restoring essential groups in the enzyme molecule to their functional state. This is the case for the activation, or re-activation, of creatine kinase by sulphydryl compounds, such as reduced glutathione, mercaptoethanol, dithiothreitol, or N-acetyl cysteine, which ensure the reduction of thiol groups necessary for catalysis. A method for estimation of the MB isoenzyme of creatine kinase in human serum has been based on differential activation of this isoenzyme and the MM dimer by a pair of sulphydryl reagents, glutathione and dithiothreitol.[55] Serum samples are divided into two portions, to one of which dithiothreitol is added to a concentration of

12.34 mmoles l^{-1}. The creatine kinase activities of the two samples are then determined with reagents in which reduced glutathione (8.7 mmoles l^{-1}) is present. The activity in the sample unsupplemented with dithiothreitol represents that of the MM isoenzyme alone, whereas in the supplemented sample it is the sum of both MM and MB isoenzyme activities. Good correlation of estimates of MB-creatine kinase by this method with values obtained by quantitative zone electrophoresis has been reported, with a coefficient of variation less than $\pm 10\%$ for elevated activities of the isoenzyme.[55]

The molecular basis of the selective-activation method is not clear. Glutathione is apparently incapable of activating the MB isoenzyme when the latter is present in mixtures with MM-creatine kinase, although this reducing agent will activate the isolated MB dimer; to achieve full activation of mixtures of the isoenzymes, the more powerful reagent, dithiothreitol, is necessary. However, the method has been widely adopted for clinical use, especially since it is adaptable to automated analysis,[56] but several reports have indicated rather poor correlation with chromatographic, electrophoretic, or immunochemical methods of determination of MB-creatine kinase.[57, 58, 59] Some of the difficulties with the selective-activation method may arise from the presence in certain samples of irreversibly inactivated isoenzymes. Furthermore, some increase in the activities of non-MB isoenzymes, including the MM form, may result from the addition of dithiothreitol, so that interpretation of the increment of activity resulting from the presence of this activator as due solely to MB-creatine kinase may cause the contribution of the latter to be overestimated. The original method was intended to be applied to the analysis of two-component, MM and MB, isoenzyme mixtures and the possible effects of the activators on BB-creatine kinase was not considered, although this isoenzyme is now known to be present in serum more frequently than was at one time believed.

Isoenzymes frequently differ in their pH optima, and the observation of irregularly shaped curves for the dependence of reaction velocity on pH can be a valuable indication of the heterogeneity of an enzyme preparation. For example, acid phosphatases from erythrocytes and prostate differ in their pH optima, and so, to a lesser extent, do the LD_1 and LD_5 isoenzymes of lactate dehydrogenase under certain conditions. However, observed pH optima are markedly affected by such factors as the nature and ionic strength of the buffer solution, the type of substrate when this can be varied, and even in some cases on the substrate concentration. Some differential effects of pH may be related to irreversible inactivation of particular enzyme forms, rather than to differences in reversible ionisations, and so may be time dependent. Careful standardisation and control of experimental conditions are necessary, therefore, if differences in pH dependence are to form a reliable basis for isoenzyme characterisation.

Significance of Catalytic Differences in Isoenzyme Analysis

Differences in catalytic properties between multiple forms of enzymes provide evidence of what may in many cases be subtle variations in the structures of ligand-binding sites, resulting perhaps from differences in primary structure near these sites or in conformation of the enzyme molecules themselves. The functional differences thus produced may provide indications of the specific biological roles of commonly occurring isoenzymes, or explain the adverse consequences of the occurrence of isoenzymes determined by rare alleles. The more marked differences in catalytic properties between isoenzymes offer the possibility of quantitative analysis of mixtures, particularly when only two components are present. Characterisation by catalytic differences is most valuable, and may even be indispensable, when the isoenzymes in question cannot readily be physically separated.

References for Chapter 5

1 Tedesco, T.A., *J. Biol. Chem.*, 1972, **247**, 6631.
2 Sutton, H.E., and Wagner, R.P., *Ann. Rev. Hum. Genet.*, 1975, **9**, 187.
3 Plagemann, P.G.W., Gregory, K.F., and Wroblewski, F., *J. Biol. Chem.*, 1960, **235**, 2288.
4 Wilson, A.C., Cahn, R.D., and Kaplan, N.O., *Nature*, 1963, **197**, 331.
5 Vesell, E.S., and Pool, P.E., *Proc. Natl. Acad. Sci. U.S.A.*, 1966, **55**, 756.
6 Eppenberger, H.M., Dawson, D.M., and Kaplan, N.O., *J. Biol. Chem.*, 1967, **242**, 204.
7 Witteveen, S.A.G.J., Sobel, B.E., and DeLuca, M., *Proc. Natl. Acad. Sci. U.S.A.*, 1974, **74**, 1384.
8 Szasz, G., and Gruber, W., *Clin. Chem.*, 1978, **24**, 245.
9 Horecker, B.L., in 'Isozymes, I Molecular Structure', ed. Markert, C.L., Academic Press, New York, 1975, p. 11.
10 Fleisher, G.A., Potter, C.S., Wakim, K.G., Pankow, M., and Osborne, D., *Proc. Soc. Exp. Biol. Med.*, 1960, **103**, 229.
11 Boyd, J.W., *Biochem. J.*, 1961, **81**, 434.
12 Rej, R.D., and Vanderlinde, R.E., *Clin. Chem.*, 1975, **21**, 1585.
13 Grimm, F.C., and Doherty, D.G., *J. Biol. Chem.*, 1961, **236**, 1980.
14 Moss, D.W., Campbell, D.M., Anagnostou-Kakaras, E., and King, E.J., *Biochem. J.*, 1961, **81**, 441.
15 Moss, D.W., and King, E.J., *Biochem. J.*, 1962, **84**, 192.
16 Moss, D.W., Campbell, D.M., Anagnostou-Kakaras, E., and King, E.J., *Pure Appl. Chem.*, 1961, **3**, 397.
17 Bernstein, L.H., Everse, J., Shioura, N., and Russell, P.J., *J. Mol. Cell. Cardiol.*, 1974, **6**, 297.
18 Bernstein, L.H., *Clin. Chem.*, 1977, **23**, 1928.
19 Hunter, R.L., and Markert, C.L., *Science*, 1957, **125**, 1294.
20 Hunter, R.L., and Burstone, M.S., *J. Histochem. Cytochem.*, 1960, **8**, 58.
21 Moss, D.W., and Walli, A.K., *Biochim. Biophys. Acta*, 1969, **191**, 476.

22 Davies, D.R., *Biochem. J.*, 1934, **28**, 529.
23 Abul-Fadl, M.A.M., and King, E.J., *Biochem. J.*, 1949, **45**, 51.
24 Babson, A.L., Read, P.A., and Phillips, G.E., *Am. J. Clin. Pathol.*, 1959, **32**, 88.
25 Roy, A.V., Brower, M.E., and Hayden, J.E., *Clin. Chem.*, 1971, **17**, 1093.
26 Landau, W., and Schlamowitz, M., *Arch. Biochem. Biophys.*, 1961, **95**, 474.
27 Ahmed, Z., and King, E.J., *Biochim. Biophys. Acta*, 1960, **45**, 581.
28 Moss, D.W., Eaton, R.H., Smith, J.K., and Whitby, L.G., *Biochem. J.*, 1967, **102**, 53.
29 Eaton, R.H., and Moss, D.W., *Biochem. J.*, 1967, **105**, 1307.
30 Fishman, W.H., Green, S., and Inglis, N.I., *Biochim. Biophys. Acta*, 1962, **62**, 363.
31 Moss, D.W., *Clin. Chim. Acta*, 1971, **35**, 413.
32 Haije, W.G., *Clin. Chim. Acta*, 1973, **48**, 23.
33 Whitaker, K.B., and Moss, D.W., *Clin. Chim. Acta*, 1974, **52**, 347.
34 Rosalki, S.B., and Wilkinson, J.H., *Nature*, 1960, **188**, 1110.
35 Plummer, D.T., Elliott, B.A., Cooke, K.B., and Wilkinson, J.H., *Biochem. J.*, 1963, **87**, 416.
36 Elliott, B.A., and Wilkinson, J.H., *Lancet*, 1961, **i**, 698.
37 Konttinen, A., and Halonen, P.I., *Am. J. Cardiol.*, 1962, **10**, 525.
38 Kaplan, N.O., and Ciotti, M.M., *Ann. N.Y. Acad. Sci.*, 1961, **94**, 701.
39 Cahn, R.D., Kaplan, N.O., Levine, L., and Zwilling, E., *Science*, 1962, **136**, 962.
40 Kun, E., and Volfin, P., *Biochim. Biophys. Res. Commun.*, 1966, **22**, 187.
41 Campbell, D.M., and Moss, D.W., *Clin. Chim. Acta*, 1961, **6**, 307.
42 Pfleiderer, G., and Jeckel, D., *Biochem. Z.*, 1957, **329**, 370.
43 Wieland, T., Pfleiderer, G., and Ortanderl, F., *Biochem. Z.*, 1959, **331**, 103.
44 Plummer, D.T., and Wilkinson, J.H., *Biochem. J.*, 1963, **87**, 423.
45 Bodansky, O., *J. Biol. Chem.*, 1937, **118**, 341.
46 Kellen, J.A., and Lustig, V., *Oncology*, 1971, **25**, 239.
47 Fishman, W.H., and Sie, H.-G., *Enzymologia*, 1971, **41**, 141.
48 Doellgast, G.J., and Fishman, W.H., *Nature*, 1976, **259**, 49.
49 Borgers, M., *J. Histochem. Cytochem.*, 1973, **21**, 812.
50 Van Belle, H., *Clin. Chem.*, 1976, **22**, 972.
51 Fishman, W.H., Inglis, N.I., and Krant, M.J., *Clin. Chim. Acta*, 1965, **12**, 298.
52 Tan It-Koon, and Moss, D.W., *Clin. Chim. Acta*, 1969, **25**, 117.
53 Kalow, W., and Genest, K., *Can. J. Biochem. Physiol.*, 1957, **35**, 339.
54 Harris, H., and Whittaker, M., *Nature*, 1961, **191**, 496.
55 Rao, P.S., Lukes, J.J., Ayres, S.M., and Mueller, H., *Clin. Chem.*, 1975, **21**, 1612.
56 Bostick, W.D., and Mrochek, J.E., *Clin. Chem.*, 1977, **23**, 1633.
57 Balkcom, R.M., *Clin. Chem.*, 1976, **22**, 926.
58 Vacca, G., *Clin. Chim. Acta*, 1977, **75**, 175.
59 Morin, L.G., *Clin. Chem.*, 1977, **23**, 205.

6

Methods of Obtaining Structural Information

A full understanding of the structural diversity underlying the existence of isoenzymes can only be achieved by determination of the amino-acid sequences of their respective polypeptide chains, and of the conformations and associations of the one or more chains which together constitute the isoenzymic molecules. Additional information as to the existence and nature of interactions between enzyme molecules and non-protein or non-enzymic components may also be required before enzyme heterogeneity arising at the post-genetic level can be explained. Complete information on comparative structures of isoenzymes is still relatively scanty. Differences in amino-acid composition have been reported between the members of several families of isoenzymes, but the corresponding primary structures have been established in a few cases only; e.g., for human erythrocyte carbonic anhydrases B and C (isoenzymes I and II),[1] and mitochondrial and cytoplasmic aspartate aminotransferases from pig heart.[2,3] Three-dimensional structures of crystalline complexes of the lactate dehydrogenase isoenzymes 1 and 5, from pig heart and dogfish muscle, respectively, have been compared by X-ray crystallography at 6 Å resolution[4] and those of B and C carbonic anhydrases[5] at a resolution of 2 Å. Although the number of such fully definitive structural studies will continue to grow, the difficulties of adding to them are considerable, not only because of the laboriousness of the experimental methods but also because of the requirement for significant quantities of pure isoenzymes including, for X-ray crystallography, native and heavy-atom substituted crystalline forms. Much attention has therefore been directed towards methods which, while falling short of complete definition of structural differences, nevertheless allow some inferences to be drawn about probable variations in structure between multiple enzyme forms. Some of these approaches are considered in this chapter.

Quaternary Structure of Isoenzymes

Probably the majority of enzyme molecules consist of aggregates of smaller subunits or monomers, each consisting of a single polypeptide chain; *i.e.*, the molecules possess a quaternary level of structure. The association of identical and non-identical subunits in various com-

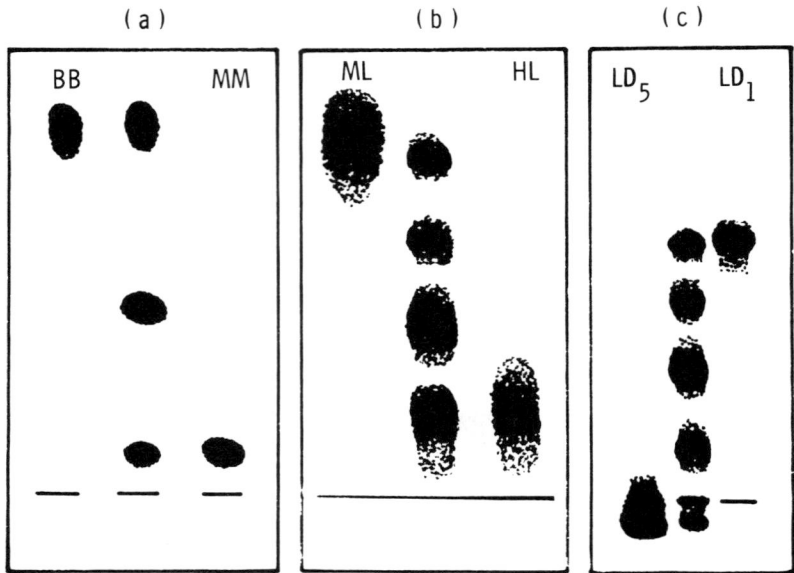

Figure 6.1   *Diagrams of the results of experiments on the formation* in vitro *of hybrid isoenzymes*
   (a) *Rabbit muscle* (MM) *and brain* (BB) *isoenzymes of creatine kinase hybridised by mixing in a solution of urea* (8 moles l^{-1}) *and β-mercaptoethanol* (0.1 mole l^{-1}) *at pH* 7.5, *followed by dialysis against buffer solution containing mercaptoethanol* (based on data from D. M. Dawson, H. M. Eppenberger, and N. O. Kaplan, *J. Biol. Chem.*, 1967, **242**, 210)
   (b) *Mouse liver* (ML) *and human liver* (HL) *isoenzymes of nucleoside phosphorylase mixed and frozen and thawed in a solution of* 2 moles l^{-1} NaCl (adapted from Y. H. Edwards, D. A. Hopkinson, and H. Harris, reference 16)
   (c) *Purified ox-heart lactate dehydrogenase isoenzymes* LD_1 *and* LD_5 *hybridised by freezing and thawing in* NaCl *solution* (adapted from C. L. Markert, reference 6)
   In each case the isoenzyme mixtures were separated by starch-gel electrophoresis (anode at the top) *with the hybridised sample in the centre. The formation is demonstrated of one hybrid isoenzyme of dimeric creatine kinase, two of trimeric nucleoside phosphorylase, and three of the tetrameric lactate dehydrogenase*

binations to form catalytically active molecules is a frequent cause of enzyme polymorphism, and methods for investigating this phenomenon are correspondingly important. Dissociation of polymeric protein molecules into their component monomers with the aid of reagents such as urea or guanidine hydrochloride, with in some cases the additional use of reagents which break disulphide bridges, is an established procedure in the structural analysis of proteins. Some techniques for determining the sizes of the intact polymer and its protomers have been mentioned in Chapter 2; *e.g.*, gel-filtration chromatography, gradient-pore electrophoresis, and electrophoresis in the presence of dodecyl sulphate. However, the subunits of isoenzyme molecules thus produced are usually catalytically inactive, so that the parent isoenzymes must be available in quantities and states of purity which allow the detection and identification of the dissociated subunits by methods independent of enzymic activity. These limitations are avoided when dissociation of one or more types of isoenzyme molecules is followed by a reversal of the conditions so as to permit re-association of the protomers into new polymeric combinations, each with catalytic activity, from the number and properties of which the probable quaternary structures of the original polymers can be inferred. For example, two different types of subunits can combine to form five possible tetrameric isoenzymes, four if the active forms are trimers and three in the case of dimeric enzymes (Fig. 6.1). The properties of the hybrid isoenzymes are usually intermediate between those of the respective homopolymeric forms. The ability of isoenzyme subunits from different tissues or species to associate in forming active polymers also provides evidence of conservation of polypeptide structures during the course of evolution of different species, or of structural similarities between polypeptides determined by multilocular or allelic genes in a single species. These hybridisation experiments, as they are called, can be carried out in various ways.

Hybridisation *in vitro*.—When a mixture of isoenzymes 1 and 5 of lactate dehydrogenase in equal proportions is frozen in 1 mol l^{-1} sodium chloride solution and then thawed, the two homopolymers dissociate into their component H- and M-subunits which re-combine to produce the five possible tetramers (H_4, H_3M, H_2M_2, HM_3, and M_4) in the proportions expected for random re-association, separable by zone electrophoresis.[6] This type of experiment has been widely applied and extended, not only to the production of inter- and intra-species hybrid molecules of lactate dehydrogenase,[7,8] but also to many other isoenzyme systems. Although mixtures of the homopolymeric isoenzymes are often used as sources of the different subunits, dissociation of a single type of heteropolymer with subsequent generation of new isoenzymic combinations has also been demonstrated. Tetrameric structures have been inferred for aldolase,[9] malic enzyme (EC 1.1.1.40),[10] phosphofructokinase (EC 2.7.1.11),[11] and pyruvate kinase[12] by *in vitro* hybridisation. Isoenzymes shown in this way to be dimers include liver alcohol dehydrogenase,[13] glucose-6-phosphate dehydrogenase,[14] and creatine kinase,[15] and a trimeric structure has been demonstrated for nucleoside phosphorylase (EC 2.4.2.1).[16] Hybridisation of the subunits which compose the isoenzymes of human hexosaminidase has also been achieved,[17] although some uncertainty still exists as to whether the active molecules are dimers or tetramers.[18, 23]

Although mild conditions such as freezing and thawing in solutions of high ionic strength induce dissociation and recombination in many cases, more vigorous treatment may be necessary; *e.g.*, inter- and intra-species hybridisation of bovine and chicken pyruvate kinases required, first, dissociation with guanidine hydrochloride, followed by re-naturation by dilution into buffer with dithiothreitol or β-mercaptoethanol,[19] and treatment with 8 mol l^{-1} urea solution for 30 min with subsequent removal of urea by dialysis in the presence of β-mercaptoethanol was needed to produce hybrid creatine kinase molecules containing two forms of the M-subunit.[20] Dissociation and re-combination during electrophoresis has been effected in the case of isoenzymes of the cytoplasmic form of superoxide dismutase (EC 1.15.1.1) by carrying out the electrophoretic separation in starch gel at a temperature of 40 °C.[21] At lower temperatures, extracts of liver tissue from subjects heterozygous for the enzyme show the three-banded pattern expected for a dimeric enzyme such as this. At the higher temperature, however, only zones corresponding to the homopolymers are seen. This is explained by the existence of an equilibrium between the dimeric isoenzymes and their constituent monomers above about 35 °C. Since the two types of monomer are undergoing separation due to their different net charges during electrophoresis, re-formation of the homodimers can occur, but not of mixed dimers.

In these applications of the *in vitro* hybridisation technique, the different subunits which are induced to dissociate and re-combine are the naturally occurring products of separate gene loci or alleles. However, in one case in which no natural isoenzyme with electrophoretic properties distinct from those of the isoenzyme under study existed, with which hybridisation experiments could be carried out, enzymic modification was used to produce a set of subunits with altered net charge. This study concerned the C_4 component of cholinesterase, the major zone of this enzyme seen after electrophoresis of serum in starch gel.[22] The isoenzyme purified from human plasma was treated with neuraminidase to reduce its net negative charge at alkaline pH by removal of terminal *N*-acetyl neuraminic acid residues. Hybridisation of the modified isoenzyme with its native counterpart was achieved by freezing and thawing a mixture of the two 3 times in a solution of sodium chloride (4 mol l^{-1}) and mercaptoethanol (1% m/V) at pH 10—11, then keeping the mixture at 4 °C for at least 24 h. Subsequent electrophoresis showed the unmodified and modified isoenzymes together with three new components migrating between them, a result consistent with a tetrameric structure of the native cholinesterase isoenzyme.

Hybridisation in Living Cells.—Hybrid isoenzymes occur naturally in cells in which two or more gene loci or alleles at a single locus are active in the production of different enzyme subunits (Fig. 1.3) and comparisons of the isoenzyme patterns of extracts of different tissues, particularly of the patterns presented by homozygotic and heterozygotic individuals, have provided a fruitful source of insights into the nature and extent of this aspect of enzyme polymorphism. An opportunity to extend the analysis of hybrid isoenzyme formation in living cells beyond the limits of observation of naturally occurring examples into the domain of experimentation has been provided by the development of techniques of somatic-cell hybridisation and culture. Fusion of somatic cells from different species can be induced to take place in the presence of chemical or viral agents, such as polyethylene glycol or inactivated Sendai virus, to form cell lines with nuclei which contain functional chromosomes from both parent cells. The hybrid cells usually retain the full chromosome complement of one cell, although part of the genome of the other is lost; thus, analogous genes derived from both parent cells are expressed to different extents in the daughter cells and their respective products can be detected in those cases in which differences in properties exist. Differences in electrophoretic mobility have been established for more than 60 enzymes and isoenzymes derived either from human genes or from the genes of mouse or Chinese hamster cells,[23] so that the patterns of enzyme zones obtained by electrophoresis of extracts of human cells which have been fused with cells from one of the other species reflect the genetic composition of the hybrid cells.

The retention or loss of particular enzymes by different somatic-cell hybrids, and the association of these events with the appearance or disappearance of other chromosomal markers, enable the identity of the chromosomes which carry the enzyme-determining genes to be deduced. Human chromosomes, rather than those of rodent origin, tend to be lost during fusion of human and rodent cells, so that the technique can be used to map the human chromosomes. More than 75 gene loci determining human enzymes and isoenzymes have now been assigned to chromosomes in this way.[23] Information of this kind is valuable in illuminating the probable evolution of metabolic pathways and in understanding the origins of inherited metabolic diseases. However, some limitation on the scope of the method is imposed by the tendency of enzyme activities associated with the differentiation of cells to be lost during culture, although enzymes catalysing the more fundamental metabolic processes continue to be expressed.

The expression of genes determining different isoenzyme subunits in hybrid somatic cells leads to the appearance in these cells of heteropolymeric isoenzymes, when the different subunits retain the structural features necessary for polymerisation. The cell-fusion technique thus supplements and extends the experimental production of hybrid isoenzymes *in vitro*, and the appearance of electrophoretically separated isoenzyme zones from hybrid cells is open to similar interpretations as to the subunit composition of the respective isoenzymes. The quaternary structures of nearly 50 enzymes and isoenzymes have already been confirmed or determined by the analysis of isoenzyme patterns of hybrid cells.[23, 24]

Secondary and Tertiary Structures

These levels of structure of protein molecules refer to the three-dimensional conformations assumed by short lengths of the individual constituent polypeptide chains and by the whole chains. Secondary and tertiary structures are themselves apparently determined by the linear sequence of amino acids in a particular polypeptide chain, *i.e.*, by its primary structure, since the chain tends to assume its most stable configuration. Within the more restricted usage of the term, isoenzymes are the products of distinct structural genes and consequently differ in the primary structures of their polypeptide chains; therefore, differences between isoenzymes in secondary and tertiary structures are also to be expected, although considerable similarities will exist when variations between the respective amino-acid sequences are few. Maintenance of the three-dimensional structures of protein molecules depends mainly on the large numbers of hydrogen bonds and hydrophobic interactions which exist between adjacent amino-acid side chains, together with covalent linkages such as disulphide bridges in some cases. Differences between isoenzymes in their resistance to denaturing agents (Chapter 3) reflect variations in the number and strength of these stabilising interactions, thus providing what is probably the most readily obtainable evidence of conformational differences. However, such observations cannot be interpreted in terms of specific structures.

When samples of purified isoenzymes are available, physical techniques can be used to deduce the presence of three-dimensional structures such as α-helices. Measurement of optical rotatory dispersion for three chromatographically separated forms of mitochondrial aspartate aminotransferase from pig heart indicated that these forms have similar conformations.[25] This method was also used to investigate postulated differences in conformation amongst sub-forms of malate dehydrogenase from chicken mitochondria, and to compare the native and iodinated forms.[26] The results suggested that, of the naturally occurring forms, the one which is most cathodal on electrophoresis possesses the greatest content of helical structures. The introduction of between approximately 1 and 3 atoms per mole of iodine into this molecular species produced differences in optical rotatory dispersion comparable to those observed amongst the less-cathodal native forms, possibly due to partial unfolding of secondary and tertiary structures. Iodination also resulted in the appearance of forms with electrophoretic mobilities similar to those of the native variants.

These observations with respect to mitochondrial malate dehydrogenase gave rise to the hypothesis that conformational isomerism might be a more general cause of enzyme heterogeneity (Chapter 1, p. 11; Fig. 1.4), and modifications in electrophoretic mobility of multiple forms of other enzymes, produced by treatment with potentially denaturing agents or by the binding of ligands, have also been interpreted in these terms. Examples include the altered mobilities of isoenzymes of alcohol dehydrogenase from horse liver or *Drosophila melanogaster* in the presence of its coenzyme, nicotinamide-adenine dinucleotide.[27, 28] However, the experimental evidence is not fully consistent with the conformer hypothesis, for either malate dehydrogenase[29] or other enzymes, and other explanations cannot be excluded; *e.g.*, variations in quaternary structure due to the existence of subunits with only minor differences between them, or alterations in net charge due to ligand-binding.

Specific ligands which become attached to specialised regions of enzyme molecules themselves offer a means of exploring the topography of these sites, and topographical differences between analogous regions of isoenzymic molecules. Differences in the properties of binding sites for substrates, cofactors or inhibitors underlie many of the variations between isoenzymes in their catalytic behaviour discussed in the previous chapter and, although some of these inter-isoenzymic variations are due to the presence of different chemical groupings, others must derive from differences in the three-dimensional structures of active sites. Thus, the exploration of ligand-binding sites involves consideration of secondary and tertiary structures, as well as primary structure. A particular structural feature is common to enzymes which require nucleoside phosphates as coenzymes or substrates. This is the 'dinucleotide fold', a zone of secondary and tertiary structure forming non-polar pockets into which the aromatic rings of the nucleotides fit. The dye Cibachron Blue F3GA has structural resemblances to nucleotides and can therefore become bound to the dinucleotide fold of enzymes and isoenzymes which possess this feature. Attachment of the dye is accompanied by a spectral change which can be detected by difference spectrophotometry. Dissimilar dissociation constants for the binding of Cibachron Blue by the M_4 and H_4 isoenzymes of lactate dehydrogenase, from rabbit muscle and beef heart, respectively, have been reported,[30] suggesting that the topography of the dinucleotide fold is not identical in these two isoenzymes.

Primary and Other Covalent Structures

Peptide Maps and Partial Sequences.—The most sensitive and specific indication of differences in primary structures of closely similar proteins such as isoenzymes, short of determination of complete amino-acid sequences, is given by comparison of two-dimensional maps of peptides obtained by partial hydrolysis of the proteins with enzymes or acid. This

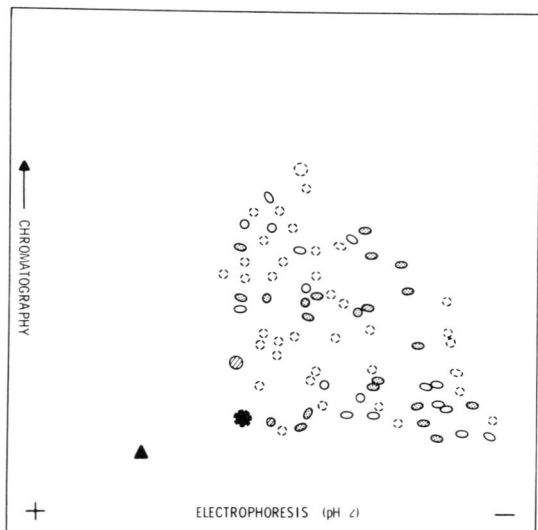

Figure 6.2 *Two-dimensional separation on a thin-layer cellulose plate, by high-voltage electrophoresis at pH 2 followed by chromatography in n-butanol:pyridine:acetic acid:water (21:12:2:15 V/V), of a mixture of peptides obtained by tryptic digestion of human placental alkaline phosphatase labelled with ^{32}P. The radioactive peptide (solid) derived from the active centre of the isoenzyme is readily identified amongst the large number of peptides derived from other regions of the molecule or from possible protein impurities, stained with fluorescamine. The point of application is indicated by the triangle. [Reproduced, with permission, from K. B. Whitaker, P. G. H. Byfield and D. W. Moss, Clin. Chim. Acta (Elsevier), 1976, 71, 285]*

'fingerprinting' technique has been shown to be capable of demonstrating the single amino-acid substitutions which are, in many cases, the only primary structural differences between the products of allelic genes. Examples of its applications to isoenzyme analysis include studies of the isoenzymes of lactate dehydrogenase,[31] alkaline phosphatase,[42] and creatine kinase,[32] and comparison of the normal form of the latter enzyme with a variant present in muscle of dystrophic mice.[33] Identical maps were obtained from the multiple forms of mitochondrial aspartate aminotransferase.[25] Cleavage with enzymes such as trypsin or pepsin is preferable to partial acid or alkaline hydrolysis because the specificity of proteolytic enzymes ensures that their action is reproducible and non-random. However, long periods of digestion with enzymes or even an initial partial denaturation by heat, acid, or other agents may be required, since some isoenzyme proteins are markedly resistant to proteolysis. Separation of peptides from undegraded protein by gel filtration is a useful preliminary to separation by high-voltage electrophoresis or chromatography, or both. Plates coated with thin layers of alumina or cellulose for two-dimensional separations, or polyacrylamide gels for unidimensional electrophoresis, have now generally replaced the filter-paper of earlier studies as supporting media for resolution of the peptide mixtures.

Comparison of maps of peptides stained with detection reagents such as ninhydrin or fluorescamine assumes that each isoenzyme is free of other proteins, a condition that is difficult to satisfy when abundant sources of the isoenzymes are not available; *e.g.*, as is the case for variant isoenzymes determined by allelic human genes. The specificity of the method can be improved by attaching a label to the isoenzyme molecules before partial hydrolysis, the presence of which subsequently identifies those peptides containing it as having been derived from the isoenzyme and not from protein impurities (Fig. 6.2). Several labels have been used in this way, selected for their known or presumed affinities for particular active sites. The specificity of combination between the label and the isoenzyme can be further verified by demonstrating reduced binding in the presence of the substrate or a competitive inhibitor.

The presence of pyridoxal phosphate as a prosthetic group at the active centre of aminotransferases provides a means of identifying peptides derived from this molecular region. Radioactive peptides containing pyridoxal phosphate linked to lysine have been obtained by reducing mitochondrial aspartate aminotransferase with tritiated lithium borohydride, then digesting the isoenzyme with the proteolytic enzyme thermolysin.[35] The amino-acid sequence of the radioactive peptide was identical to that of peptides from corresponding isoenzymes from the mitochondria of other species. Active-centre peptides from cytoplasmic aspartate aminotransferases show inter-species identity, but with some differences between the cytoplasmic and mitochondrial isoenzymes. Radioactive peptides derived from the active centres of various alkaline phosphatases have been obtained by partial hydrolysis of these enzymes after labelling with ^{32}P-orthophosphate. At acid pH, alkaline phosphatases incorporate orthophosphate rapidly and irreversibly, with formation of phosphorylserine. The covalent bond thus formed survives hydrolysis of the enzyme protein by proteolytic enzymes or acid, and radioactive peptides have been isolated from alkaline phosphatases from micro-organisms, animal tissues, and human placenta[36, 37, 38] (Fig. 6.2).

Marked resemblances exist, in electrophoretic mobility, in amino-acid composition and even in partial sequences, between active-site peptides from alkaline phosphatases of evolutionarily remote species (Table 6.1)

Table 6.1 *Amino-acid compositions (residues/mole) of active-centre peptides from* E. coli *alkaline phosphatase[36] and from human placental alkaline phosphatase,[38] obtained by partial digestion of the enzymes after labelling with* [32]P

Amino acids present in placental peptide	Amino acids present in E. coli *peptide*
Ser P	Ser P
Ala × 3	Ala × 3
Asp[a] × 2	Asp × 2
Thr × 2	Thr × 2
Gly	Gly
Pro	Pro
Ser	Ser
Val	Val
Arg	Lys
Glx × 2	Tyr
Gly	
His	
Leu	

[a] May be asparagine in peptide from placental phosphatase

as is the case also with the aminotransferase isoenzymes. This supports the view that the structure of the active-centre region of enzyme molecules is strongly conserved during evolution and mutation and points to a probable limitation of the principle of comparing peptides derived from these regions of different isoenzymes. Unless the different peptides are large enough to encompass portions of the primary structure which are relatively remote from the active site, it is perhaps less likely that differences in amino-acid sequence will be found between peptides from isoenzymes which are functionally very similar. However, preliminary results indicate that the radioactive peptide separated after tryptic digestion of human kidney alkaline phosphatase labelled with ^{32}P-orthophosphate is different from that obtained in a similar way from the placental isoenzyme (K. B. Whitaker and D. W. Moss; unpublished). Isoenzymes with altered catalytic properties (and therefore presumably with structural differences at the active site) produced by allelic mutation may also prove to be exceptions to this generalisation; different patterns of radioactive peptides were obtained after electrophoresis of partial hydrolysates of the usual and atypical allelozymes of human serum cholinesterase, after these isoenzymes had been labelled with radioactive di-isopropyl fluorophosphonate.[39]

Regions of isoenzyme molecules more remote from their active centres may be accessible to study by the identification of carboxy- or amino-terminal amino-acid residues, and possibly of short sequences of amino acids adjacent to these residues, when purified preparations are available. The C-terminal sequences of carbonic anhydrase isoenzymes were shown to differ by successive removal of amino-acid residues with carboxy-peptidase.[40] Identical N-terminal sequences, extending to four residues, were found for normal human placental alkaline phosphatase and a tumour-derived variant, but these differed from the corresponding sequence for liver alkaline phosphatase.[41,42] Stepwise removal of N-terminal amino acids from component peptides of enzyme molecules, *e.g.*, by the Edman degradation, can usually now be repeated for about 10—20 residues with the aid of techniques in which the peptide is attached to a solid phase.

Selective Chemical or Enzymic Modification.—Many of the labels useful in isolating active-site peptides can also be employed to identify particular groups which form part of the active centres of isoenzymes; *e.g.*, di-isopropyl fluorophosphonate in the case of cholinesterase and other esterases, or radioactive orthophosphate in the study of alkaline phosphatases. However, since differences between isoenzymes in the characteristics of their active centres are generally quantitative rather than qualitative, these specific ligands find their place in isoenzyme analysis mainly in measurements of relative affinities of isoenzymes for them (Chapter 5), rather than in the exploration of the topography of binding sites. As with the technique of peptide mapping, methods which give access to parts of the isoenzyme molecules further from their active centres may be more likely to reveal the nature of inter-isoenzymic differences. Several chemical or enzymic modification procedures can provide evidence of the presence in isoenzyme molecules of particular chemical groupings or linkages. In some cases, the altered molecules thus produced resemble the multiple forms of an enzyme found in extracts of cells or tissues, suggesting ways in which post-genetic modifications may occur *in vivo.*

The most useful modifications of isoenzyme molecules are those which result in some marked change in properties while preserving catalytic activity, since this permits the study of the effects of modification even when only impure or limited isoenzyme samples are available. Methods which produce changes in net molecular charge are particularly advantageous because of the altered electrophoretic mobilities of catalytically active isoenzyme zones. The net charge of protein molecules is determined by the numbers and states of ionisation of several types of amino-acid side-chains in contact with the aqueous environment;

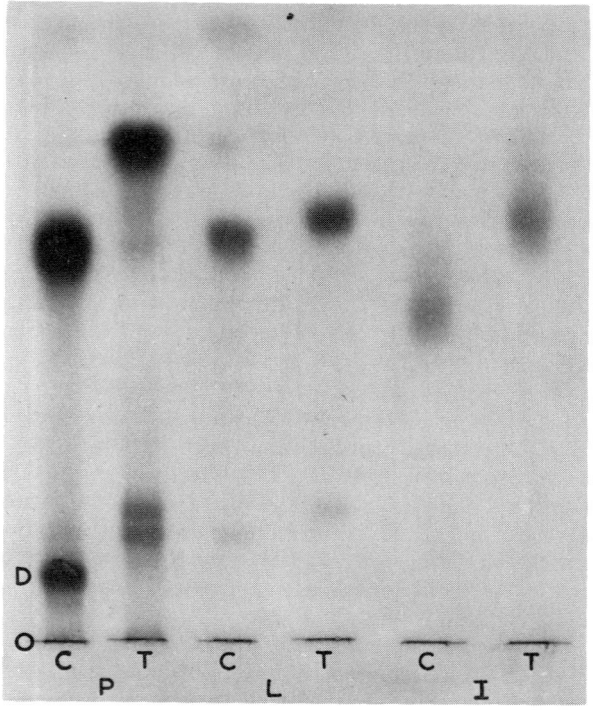

Figure 6.3   *Effect of acetylation on the electrophoretic mobilities of human placental (P), liver (L) and small-intestinal (I) alkaline phosphatases on starch-gel electrophoresis. C, control samples; T, treated samples; O, origin. D is a minor zone of placental alkaline phosphatase which is resolved into two by acetylation. The anode is at the top. Active enzyme zones were stained as in Fig.* 2.3. [Reproduced, with permission, from: D. W. Moss, *Enzymologia* (W. Junk), 1970, **39,** 319]

principally the amino- and carboxyl-groups of lysine, hydroxylysine, and arginine, and aspartic and glutamic acids, with, to a lesser extent, the hydroxyl-groups of serine, threonine, and tyrosine, the imidazole ring of histidine, and the sulphydryl group of cysteine. The ionisation of these groups can be modified or prevented by chemical treatments of varying degrees of specificity, so that selective effects of such treatments on the electrophoretic mobilities of individual isoenzymes can be interpreted in terms of the probable relative numbers of the modifiable residues present in the molecules of the respective isoenzymes. Examples of modification procedures include esterification of carboxyl-groups, acylation of amino-groups with acetic or succinic anhydrides, nitration of tyrosine residues with tetranitromethane, and reaction of sulphydryl-groups with reagents such as N-ethyl maleimide. Some loss of enzymic activity may result from non-specific denaturation during these treatments, although the necessary conditions are usually mild. With some enzymes, groups essential for substrate binding or conversion may be modified, with loss of activity, although this may be prevented to some extent by carrying out the reaction in the presence of the substrate or a competitive inhibitor.

Examples of the use of chemical modification and its effects on electrophoretic mobility can be found in studies of human alkaline phosphatase isoenzymes.[43,44] Acetylation at 0 °C with acetic anhydride (100 mmol l^{-1}) in half-saturated ammonium sulphate solution at pH 8 reduces the activities of alkaline phosphatases from human placenta, liver, and small intestine by as much as 75% in some experiments; however, the catalytic properties of the remaining active molecules are in general similar to those of the untreated isoenzymes, suggesting that this loss of activity is due to non-specific denaturation, rather than to modification of catalytically important groups. The anodal mobility of all three modified, active isoenzymes on starch-gel electrophoresis at pH 8.6 is increased by acetylation, but this effect is less marked for liver phosphatase than for the isoenzymes from small intestine or placenta (Fig. 6.3). Since acetylation under these conditions probably selectively modifies the amino-groups of lysine and arginine, it may be inferred that these groups make a smaller contribution to the net charge of native hepatic alkaline phosphatase than is the case for the other two isoenzymes. Carbamoylation of amino-groups with sodium cyanate at pH 9.7 and 37 °C produces similar differential effects. Acetylation or succinylation of the three common allelic variants of placental alkaline phosphatase, produces a greater acceleration of the slow (S) than of either the fast (F) or intermediate (I) isoenzymes. Smaller increases in mobility are produced by treatment with tetranitromethane, a relatively specific modifier of tyrosine residues, but these are of the same order of magnitude for all three isoenzymes. Acetylation has been used to separate mixtures of electrophoretically similar isoenzymes of human placental and liver alkaline phosphatases in serum.[45]

Enzymic modification of protein structure offers the advantage of high specificity which is inherent in other uses of enzymes as analytical reagents. Human pancreatic amylase exhibits a series of zones on electrophoresis, the least-anodal fraction being the most prominent. De-amidation, with a resultant increase in the number of ionisable carboxyl-groups and thus of the net negative charge, has been suggested as the explanation of the more-anodal bands.[46] Experimental confirmation of this hypothesis has been obtained by enzymic modification of purified human pancreatic amylase with bacterial peptidoglutaminases.[47] The two peptidoglutaminases used in this work, from *B. circulans*, have somewhat different specificities; although both convert glutamine residues in polypeptides to glutamic acid, one acts on residues closer to the carboxy-terminus of chains than the other. Both produced modified pancreatic amylase zones similar to those seen when pancreatic juice itself is incubated but at rather different rates, suggesting the existence of several modifiable glutamine residues located at different sites in the amylase molecule.

Non-protein Components of Isoenzyme Structure

The presence of non-protein components is not infrequent in enzyme molecules, especially those enzymes which derive from cellular mem-branes and other morphological elements. Variations in the structures of non-protein moieties constitute an additional potential source of enzyme heterogeneity; indeed, homogeneity of glycoprotein enzyme preparations is exceptional. A full definition of the structures of such enzymes requires analysis of the carbohydrate side-chains, usually with the aid of enzymes specific for particular carbohydrate molecules and linkages, as well as of the polypeptide cores. However, as with the underlying protein structure itself, partial information about the nature of differences between the carbohydrate portions of multiple enzyme forms can be derived from specific modifying procedures which result in a selective change in properties.

The carbohydrate side-chains of glycoproteins frequently terminate in *N*-acetyl neuraminic (sialic) acid residues, which are accessible to removal by the enzyme neuraminidase, usually obtained from *Clostridium perfringens* or *Vibrio cholera*. Hydrolysis takes place under mild conditions, *e.g.*, at nearly neutral pH and, although incubation for several hours at 37 °C may be needed for complete reaction, there is usually little loss of activity on the part of the substrate isoenzyme. Catalytic properties of glycoprotein isoenzymes are little affected by this treatment. However, the strongly-acidic sialic acid residues have, when present, a marked effect on the net charge of glycoprotein molecules over a wide range of pH, so that their removal results in a considerable reduction in electrophoretic mobility towards the anode, and the effects of digestion with neuraminidase on this property of glycoprotein isoenzymes have been widely studied. Other properties which are influenced by net molecular charge, such as solubility, may also be affected (Fig. 3.9).

Human and animal tissue alkaline phosphatases are retarded on electrophoresis after neuraminidase treatment, compared with untreated controls,[48, 49] with the exception of the small-intestinal isoenzyme, which

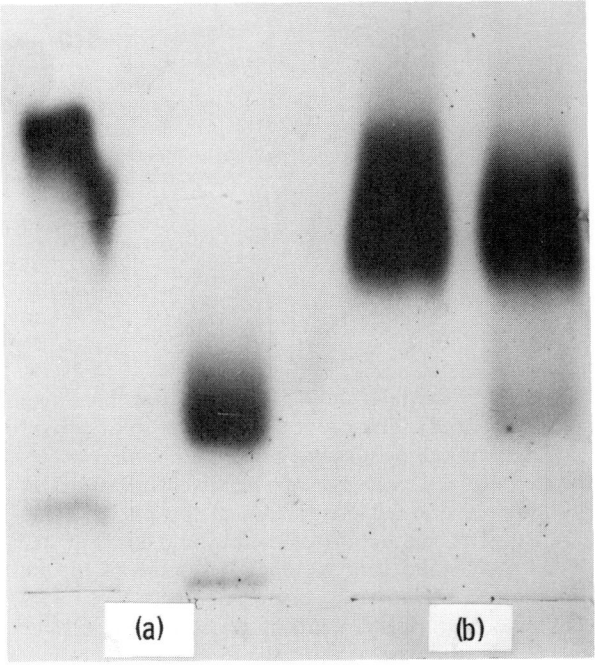

Figure 6.4 *Differential effect of digestion with neuraminidase on the mobilities of* (a) *liver and* (b) *small-intestinal alkaline phosphatases on starch-gel electrophoresis at pH 8.6. Slowly migrating zones present in the liver preparation, as well as the major zones, are retarded by this treatment, whereas the intestinal extract contains only a small proportion of neuraminidase-sensitive enzyme. Control samples are on the left and treated samples on the right of each pair, with the origin at the bottom and the anode at the top. Zones of activity were demonstrated as in Fig.* 2.3. [Reproduced, with permission, from: D. W. Moss, R. H. Eaton, J. K. Smith and L. G. Whitby, *Biochem. J.*, 1966, **98**, 32C]

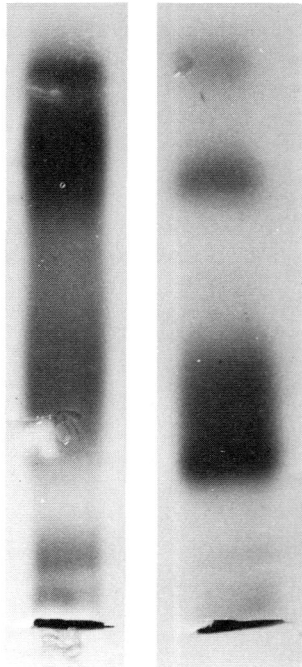

Figure 6.5 *Effect of digestion with neuraminidase on the heterogeneity of human kidney alkaline phosphatase on starch-gel electrophoresis at pH 8.6. The treated sample is on the right with a control sample to the left. The origin is at the bottom and the anode at the top. Zones of activity were demonstrated as in Fig. 2.3.* (Adapted from reference 50)

apparently contains no sialic acid residues accessible to the action of neuraminidase (Fig. 6.4). The pronounced heterogeneity with respect to net charge of human kidney alkaline phosphatase also appears to be due to a large extent to the presence of various proportions of sialic acid residues,[50] with only a small proportion of the phosphatase activity of this tissue resistant to the action of neuraminidase (Fig. 6.5). Stepwise removal of up to 7 sialic acid residues per mole of placental alkaline phosphatase has been achieved by the graded action of neuraminidase.[51] Differing contents of *N*-acetyl neuraminic acid molecules seem also to account for the many zones of prostatic acid phosphatase separable by starch-gel electrophoresis.[52]

The presence of neutral carbohydrate molecules as components of multiple forms of enzymes is less easy to determine when pure preparations are not available, since their removal may produce no marked change in properties and amounts of the carbohydrate molecules released may themselves by undetectably small.

Non-covalent attachment of entities such as non-enzymic protein, peptides, lipid or lipoprotein to enzyme proteins can give rise to multiple enzyme forms; *e.g.*, the heterogeneity of alkaline phosphatase from

Origin   Anode

(a)

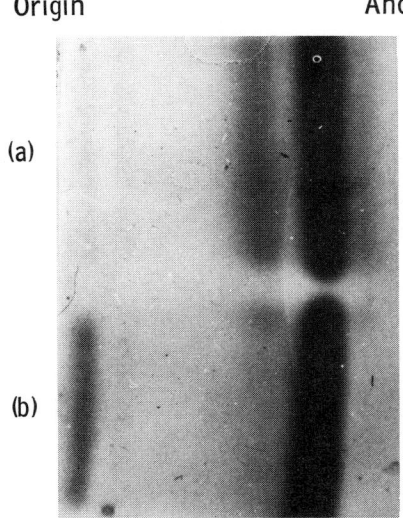

Figure 6.6 *Patterns of zones of alkaline phosphatase activity in extracts of human liver prepared by autolysis (a) and extraction with n-butanol (b) separated by starch-gel electrophoresis at pH 8.6, showing differences in the relative intensities of minor zones. Phosphatase activity was demonstrated as in Fig. 2.3.* [Reproduced, with permission, from: D. W. Moss, *Nature* (Macmillan), 1962, **193**, 981]

(b)

mouse duodenum is considerably reduced by removal of low molecular weight peptides and lipids, suggesting the existence of a common protein core to which these substances are attached.[53] Aggregation of enzyme molecules into larger units, either made up entirely of the enzyme molecules themselves or by association between enzymic and non-enzymic macromolecules, can be recognised by separation methods based on differences in molecular size (Chapter 2). This type of non-covalent interaction can sometimes be reversed by physical means, such as freezing and thawing or extraction with organic solvents, particularly when the material associated with the enzyme is more labile. These techniques have been shown to convert zones of alkaline phosphatase with low electrophoretic mobility in starch gel into smaller, more rapidly migrating forms.[54, 55]

Some of the changes in electrophoretic mobility which are occasionally seen on storage of isoenzyme preparations, or when different extraction methods are used, may be attributable to covalent or non-covalent modifications of isoenzyme molecules or molecular complexes of the kind discussed in this chapter. For example, a stepwise increase in the anodal mobility of red-cell acid phosphatase during storage at 4 °C is probably due to oxidation of sulphydryl groups,[56] while a decrease in the mobility of alkaline phosphatases may result from infection with neuraminidase-producing organisms. The different patterns of electrophoretically separable zones of liver alkaline phosphatase produced by alternative extraction procedures may represent the presence of varying amounts of macromolecular complexes[57] (Fig. 6.6). The possibility of these and similar effects must be considered when interpreting a multiplicity of zones separated by electrophoresis of enzyme preparations.

References for Chapter 6

1 Tashian, R.E., *Isozymes: Current Topics in Biological and Medical Research*, 1977, **2**, 21.
2 Doonan, S., Doonan, H.J., Hanford, R., Vernon, C.A., Walker, J.M., Bossa, F., Baron, D., Carloni, M., Fasella, P., Riva, F., and Walton, P.L., *FEBS Lett*, 1974, **38**, 229.
3 Kagamiyama, H., Sakakibara, R., Wada, H., Tamase, S., and Morino, Y., *J. Biochem.*, 1977, **82**, 291.
4 Eventoff, W., Hackert, M.L., Steindel, S.J., and Rossman, M.G., in 'Isozymes, I Molecular Structure', ed. Markert, C.L., Academic Press, New York, 1975, p. 137.
5 Notstrand, B., Vaara, I., and Kannan, K.K., in 'Isozymes, I Molecular Structure', ed. Markert, C.L., Academic Press, New York, 1975, p. 575.
6 Markert, C.L., *Science*, 1963, **140**, 1329.
7 Salthe, S.N., Chilson, O.P., and Kaplan, N.O., *Nature*, 1965, **207**, 723.
8 Zinkham, W.H., Blanco, A., and Kupchyk, L., *Science*, 1963, **142**, 1303.
9 Penhoet, E., Rajkumar, T., and Rutter, W.J., *Proc. Natl. Acad. Sci. U.S.A.*, 1966, **56**, 1275.
10 Li, J.J., *Arch. Biochem. Biophys.*, 1972, **150**, 812.
11 Tsai, M.Y., and Kemp, R.G., *Arch. Biochem. Biophys.*, 1972, **150**, 407.
12 Cardenas, J.M., and Dyson, R.D., *J. Biol. Chem.*, 1973, **248**, 6938.
13 Lutstorf, U.M., and von Wartburg, J.-P., *FEBS Lett.*, 1969, **5**, 202.
14 Yoshida, A., Steinmann, L., and Harbart, P., *Nature*, 1967, **216**, 275.
15 Dawson, D.M., Eppenberger, H.M., and Kaplan, N.O., *Biochem. Biophys. Res. Commun.*, 1965, **21**, 346.
16 Edwards, Y.H., Hopkinson, D.A., and Harris, H., *Ann. Hum. Genet.*, 1971, **34**, 395.
17 Beutler, E., and Kuhl, W., *Nature*, 1975, **258**, 262.
18 Beutler, E., Yoshida, A., Kuhl, W., and Lee, J.E.S., *Biochem. J.*, 1976, **159**, 541.
19 Cardenas, J.M., Dyson, R.D., and Strandholm, J.J., in 'Isozymes, I Molecular Structure', ed. Markert, C.L., Academic Press, New York, 1975, p. 523.
20 Wevers, R.A., Olthuis, H.P., Van Niel, J.C.C., Van Wilgenburg, M.G.M., and Soons, J.B.J., *Clin. Chim. Acta*, 1977, **75**, 377.
21 Edwards, Y.H., Hopkinson, D.A., and Harris, H., *Nature*, 1978, **271**, 84.
22 Scott, E.M., and Powers, R.F., *Nature, New Biol.*, 1972, **236**, 83.
23 Shows, T.B., *Isozymes: Current Topics in Biological and Medical Research*, 1977, **2**, 107.
24 Hopkinson, D.A., Edwards, Y.H., and Harris, H., *Ann. Hum. Genet.*, 1976, **39**, 383.

25 Michuda, C.M., and Martinez-Carrion, M., *Biochemistry*, 1969, **8**, 1095.
26 Kitto, G.B., Wassarman, P.M., and Kaplan, N.O., *Proc. Natl. Acad. Sci. U.S.A.*, 1966, **56**, 578.
27 McKinley-McKee, J.S., and Moss, D.W., *Biochem. J.*, 1965, **96**, 583.
28 Jacobson, J.B., *Science*, 1968, **159**, 324.
29 Schechter, A.N., and Epstein, C.J., *Science*, 1968, **159**, 997.
30 Thompson, S.T., and Stellwagen, E., *Proc. Natl. Acad. Sci. U.S.A.*, 1976, **73**, 361.
31 Wieland, T., Georgopoulos, D., Kampe, H., and Wachsmuth, E.D., *Biochem. Z.*, 1964, **340**, 483.
32 Dawson, D.M., Eppenberger, H.M., and Eppenberger, M.E., *Ann. N.Y. Acad. Sci.*, 1968, **151**, 616.
33 Hooton, B.T., and Watts, D.C., *Biochem. J.*, 1966, **100**, 637.
34 Saheki, S., Saheki, K., and Tanalea, T., *FEBS Lett.*, 1978, **93**, 25.
35 Bossa, F., Polidoro, G., Barra, D., Liverzani, A., and Scandurra, R., *Int. J. Pept. Protein Res.*, 1976, **8**, 499.
36 Schwartz, J.H., Crestfield, A.M., and Lipmann, F., *Proc. Natl. Acad. Sci. U.S.A.*, 1963, **49**, 722.
37 Milstein, C., *Biochem. J.*, 1964, **92**, 410; Zwaig, N., and Milstein, C., *Biochem. J.*, 1964, **92**, 421.
38 Whitaker, K.B., Byfield, P.G.H., and Moss, D.W., *Clin. Chim. Acta*, 1976, **71**, 285.
39 Muensch, H., Yoshida, A., Altland, K., Jensen, W., and Goedde, H.-W., *Am. J. Hum. Genet.*, 1978, **30**, 302.
40 Nyman, P.-O., Strid, L., and Westermark, G., *Biochim. Biophys. Acta*, 1966, **122**, 554.
41 Green, P.J., and Sussman, H.H., *Proc. Natl. Acad. Sci. U.S.A.*, 1973, **70**, 2936.
42 Badger, K.S., and Sussman, H.H., *Proc. Natl. Acad. Sci. U.S.A.*, 1976, **73**, 2201.
43 Moss, D.W., *Enzymologia*, 1970, **39**, 319.
44 Thomas, D.M., and Moss, D.W., *Enzymologia*, 1972, **42**, 65.
45 Moss, D.W., and Thomas, D.M., *Clin. Chim. Acta*, 1970, **30**, 835.
46 Karn, R.C., Rosenblum, B.B., and Merritt, A.D., *Am. J. Hum. Genet.*, 1973, **25**, 39A.
47 Ogawa, M., Kosaki, G., Matsuura, K., Fujimoto, K.-I., Minamiura, N., Yamamoto, T., and Kikuchi, M., *Clin. Chim. Acta*, 1978, **87**, 17.
48 Robinson, J.C., and Pierce, J.E., *Nature*, 1964, **204**, 472.
49 Moss, D.W., Eaton, R.H., Smith, J.K., and Whitby, L.G., *Biochem. J.*, 1966, **98**, 32C.
50 Butterworth, P.J., and Moss, D.W., *Nature*, 1966, **209**, 805.
51 Robson, E.B., and Harris, H., *Ann. Hum. Genet.*, 1966, **30**, 219.
52 Smith, J.K., and Whitby, L.G., *Biochim. Biophys. Acta*, 1968, **151**, 607.
53 Nayudu, P.R.V., and Hercus, F.B., *Biochem. J.*, 1974, **141**, 93.
54 Moss, D.W., and King, E.J., *Biochem. J.*, 1962, **84**, 192.
55 Beratis, N.G., Seegers, W., and Hirschhorn, K., *Biochem. Genet.*, 1971, **5**, 367.
56 Fisher, R.A., and Harris, H., *Ann. N.Y. Acad. Sci.*, 1969, **166**, 380.
57 Moss, D.W., *Nature*, 1962, **193**, 981.

7

Selection of Methods of Analysis

To achieve a full understanding of the nature and origin of the diversity of structures which may contribute to a particular type of enzymic activity—that is, to arrive at a complete description of each of the several enzyme forms—typically involves a progression through the various analytical stages outlined in the preceding chapters. First, the recognition of heterogeneity in the properties of an enzyme-containing extract, then the isolation and characteristion by various properties of the forms whose presence gives rise to the observed heterogeneity, and, finally, structural analysis of each of the forms. Knowledge of the nature and extent of differences in structure, together with observations on the natural occurrence of the individual forms and the possibility of their interconversion *in vitro*, permit their existence to be assigned to genetic or post-genetic causes. The rate of progress through these stages depends on many factors; the catalytic activity of the enzyme and its ease of measurement, the level of structural complexity of the enzyme molecules themselves and extent of the differences between them, and, not least, the availability of an abundant source of enzyme from which purified isoenzymes can be prepared in quantity.

Some areas of isoenzyme studies by their nature preclude the use of analytical methods which depend on the availability of considerable quantities of pure isoenzymes, so that methods such as electrophoretic fractionation or differential inactivation assume particular importance. This is particularly the case in the investigation of the rarer examples of genetically determined human enzyme polymorphism, and accounts for the great value of electrophoretic fractionation in this field. However, as has been pointed out, electrophoretic differences can arise from variations in structure other than those involving the amino-acid sequences of polypeptide chains (*e.g.*, in non-protein components) and, although both protein- and non-protein structural differences may be genetically determined, the existence of electrophoretic heterogeneity is not by itself unequivocal evidence of the existence of multi-locular or allelic genes. The biochemical geneticist therefore resorts to the observation of naturally inheritable variants of enzymes to confirm the postulated genetic origin of the variant forms which he discovers, supplemented by the artificial transfer of genetic material made possible by the technique of somatic-cell hybridisation.

The diagnostic application of isoenzyme studies does not depend on a full understanding of the molecular differences between multiple forms of enzymes, provided that tissue-specific differences in properties are retained after release of the enzyme variants into the circulation and are sufficiently marked to provide a basis for reliable and reproducible analytical methods. Two advantages are potentially present in the use of isoenzyme tests in diagnosis: the identification of specific tissues or organs as sources of a raised enzyme activity in serum, and the ability to demonstrate an abnormal activity of one or more isoenzymes, although the total enzyme activity may remain within normal limits (Fig. 1.9). The extent to which these two advantages, of improved diagnostic specificity and sensitivity, can be realised in any particular instance depends on the availability of qualitative and quantitative methods of isoenzyme analysis. It has been estimated with respect to analysis of alkaline phosphatase isoenzymes in serum that only about half of the clinically useful information potentially available is obtained by the exclusive use of qualitative methods.[1] Similar conclusions probably apply equally, or even to a greater degree, to other isoenzyme systems of diagnostic importance. For example, the main diagnostic value of the analysis of isoenzymes of creatine kinase in serum lies in the detection of minimum myocardial damage; this involves the measurement of relatively small activities of the MB isoenzyme in the presence of a large excess of the MM isoenzyme, for which quantitative methods are clearly advantageous. The respective merits of various methods of measuring creatine kinase isoenzymes have been extensively compared.[2-5] Qualitative interpretations of isoenzyme distribution are especially difficult when an elevation of serum enzyme activity derives from more than one tissue source.

Zone electrophoresis is pre-eminent amongst qualitative techniques for diagnostic applications (Figs. 1.8, 2.6). It offers the advantage of positive identification of individual isoenzyme zones, in contrast to non-separative techniques. However, its successful use imposes considerable burdens of skill and time and, in minimising these, erroneous results may be obtained through the neglect of proper control procedures. During the 20 years which have elapsed since staining reactions developed for the demonstration of enzyme activity in tissue sections were first transferred to isoenzyme electrophoresis, various precautions designed to increase the specificity of the reactions for the enzyme under study have been relaxed or abandoned. Although these simplifications are generally justified, the possibly misleading conclusions which can result from inadequately controlled separations have recently been exemplified by reports of a fluorescent artefact which may be confused with BB-creatine kinase after electrophoresis of serum, unless authentic isoenzyme markers or substrate-less blank reaction mixtures are used[6,7] (Fig. 7.1).

Electrophoresis combined with densitometric or fluorimetric scanning provides a useful method for the quantitation of well-separated isoenzyme zones, though one which is increasingly subject to error when electrophoretic resolution is incomplete. Causes of lack of proportionality between the amount of an isoenzyme applied and the area of the corresponding peak of a densitometric scan have been thoroughly

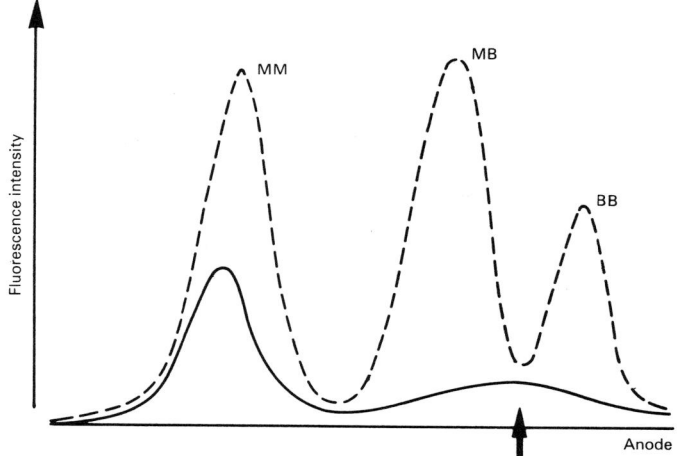

Figure 7.1 *Fluorescence scan (solid line) of serum from a patient with carcinoma of the prostate, after electrophoresis on cellulose acetate at pH 8.6 and treatment with reagents for creatine kinase demonstration. The MM isoenzyme is accompanied by a non-enzymic fluorescent zone similar in mobility to BB-creatine kinase, but not identical with it as shown by the reference preparation (broken line).* [Reproduced, with permission, from: N. Jaggarao and D. W. Moss, Clin. Chim. Acta (Elsevier), 1979, **92**, 477]

studied. However, perhaps less well appreciated are the limitations inherent in this type of method when it is necessary to measure the amount of an isoenzyme which is itself a minor component of the mixture as a whole. In such cases, the quantity of the minor isoenzyme is determined as a fraction of the total enzyme activity applied to the electrophoresis supporting medium. The contributions of all the isoenzymes, major as well as minor, must therefore be accurately measured and the activity of the predominant component must lie within the calibration range of the staining and scanning system. It is this factor which limits the total enzymic activity which can be applied, and consequently the sensitivity of the system for the minor isoenzyme component. For example, myocardial damage is accompanied by a reversal of the normal pattern of lactate dehydrogenase isoenzymes 1 and 2 in serum, so that LD_1 becomes the major component instead of LD_2, as is the case in normal serum. However, since these two components are of roughly comparable activities when the lesion is small and together account for most of the total lactate-dehydrogenase activity

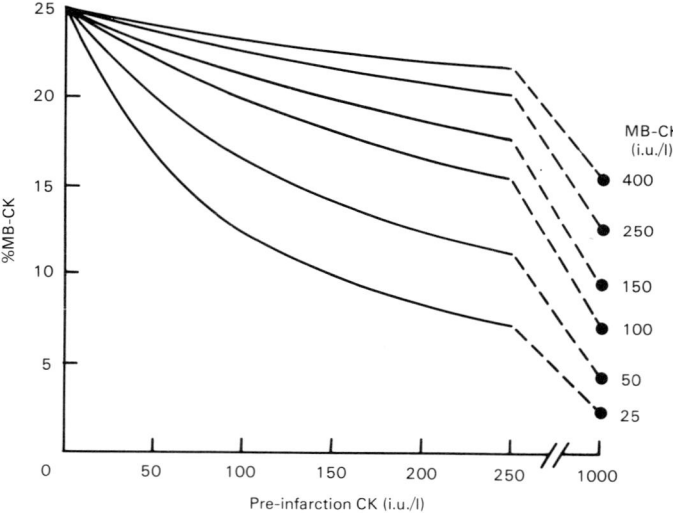

Figure 7.2   *Calculated percentages of MB-creatine kinase (CK) in serum resulting from the release of enzyme from cardiac tissue (e.g., after myocardial infarction), expressed as a function of the initial serum level of this enzyme. The effects of several magnitudes of enzyme release are shown. The simplifying assumption that the MB isoenzyme is initially absent from the serum has been made, although sensitive methods show its presence to a small extent in normal serum. MB-Creatine kinase has been assumed to constitute 25% of the total activity of creatine kinase in myocardium*

of the serum under these circumstances, quantitative electrophoresis measures both isoenzymes with similar sensitivities. A contrasting situation exists with respect to creatine kinase isoenzymes, in which an increase in the MB isoenzyme from zero to a level of 5—6% of the total creatine kinase activity may be diagnostically significant. The percentage of MB isoenzyme actually attained is influenced by factors such as the volume of cardiac tissue destroyed and the original enzyme activity in the serum (Fig. 7.2). When the latter is high due to an increased activity of MM-creatine kinase, as it generally is in the immediate post-operative period, for example, the proportion of the MB isoenzyme and the accuracy of its measurement are correspondingly reduced, compared with these parameters when an equivalent degree of myocardial-tissue damage is associated with a low initial serum enzyme activity. Elution and measurement of the activities of separated isoenzyme zones constitute an alternative quantitative electrophoresis procedure which is free from this particular limitation, and this or other methods capable of providing estimates of MB-creatine kinase which are independent of the activities of other forms of the enzyme are therefore preferable for this diagnostic application.

Some assumptions implicit in other quantitative methods of isoenzyme analysis have been pointed out in earlier chapters, such as those made in certain immunological or selective-inactivation procedures with regard to the number of different isoenzyme components which the sample is likely to contain. A combination of qualitative electrophoresis with such quantitative procedures allows the validity of these assumptions to be tested. Whatever methods of isoenzyme analysis and characterisation are chosen, strict adherence to specified experimental procedures and quality-control protocols are as important as in other branches of analysis.[8,9]

References for Chapter 7

1 Whitaker, K.B., Whitby, L.G., and Moss, D.W., in 'Enzymes in Health and Disease', ed. Goldberg, D.M., and Wilkinson, J.H., S. Karger, Basel, 1978, p. 127.
2 Morin, L.G., *Clin. Chem.*, 1977, **23**, 205.
3 Vacca, G., *Clin. Chim. Acta*, 1977, **75**, 175.
4 Fiolet, J.W.T., Willebrands, A.F., Lie, K.I., and Ter Welle, H.F., *Clin. Chim. Acta*, 1977, **80**, 23.
5 Witteveen, S.A.G.J., Hollaar, L., and Van der Laarse, A., *Clin. Chim. Acta*, 1978, **83**, 297.
6 Aleyassine, H., Tonks, D.B., and Kaye, M., *Clin. Chem.*, 1978, **24**, 492.
7 Jaggarao, N., and Moss, D.W., *Clin. Chim. Acta*, 1979, **92**, 477.
8 Rosalki, S.B., *Clin. Biochem.*, 1974, **7**, 29.
9 McKenzie, D., Clark, P.I., and Henderson, A.R., *Clin. Chem.*, 1976, **22**, 1975.